Unification in the Frame of Rotational Ether Elasticity Theory

Unification in the Frame of Rotational Ether Elasticity Theory

By

David Zareski

Cambridge Scholars Publishing

Unification in the Frame of Rotational Ether Elasticity Theory

By David Zareski

This book first published 2023

Cambridge Scholars Publishing

Lady Stephenson Library, Newcastle upon Tyne, NE6 2PA, UK

British Library Cataloguing in Publication Data
A catalogue record for this book is available from the British Library

ISBN (10): 1-5275-4003-0
ISBN (13): 978-1-5275-4003-3

TABLE OF CONTENTS

ABSTRACT

In 1920 Einstein wrote (Cf. Sec. I.1):

> *space without ether is unthinkable; for in such space there not only would be no propagation of light, but also no possibility of existence for standards of space and time.*

A very important deduction that we can make from this quotation is that **the existence of the ether permits us to precisely define the notions of time and of space.** These notions are developed further below.

As shown in our previous publications and in the following, a homogeneous elastic medium is generally governed by the Navier-Stokes-Durand equation. E. Durand completed this theory by introducing the concept of **densities of couples of force**s applied to a specific elastic medium, denoted the "ether". This medium and these densities of couples of forces had not been taken into account by them, nor, more generally, by other authors before this time. Indeed, they tended to consider only the linear forces applied to this medium. It is important to state that this concept did not permit any unification of physics, however, the Durand completion, presented in my previous publications and here, is of fundamental importance regarding this unification.

In general, our publications lead to the idea that the theory of elasticity, in which the medium is shown to be an elastic medium governed by the "appropriated Navier-Stokes-Durand equation" (ANSDE), unifies a number of theories in physics. By the word "appropriated" I mean that the ANSDE, being a general equation of elasticity, is restricted to those cases containing the fundamental elements that unify the different domains of physics. That is, the ANSDE ensues from the general theory of elasticity, with the difference that I consider that the points of this medium do not move linearly, one relative to the other, but can only rotate on themselves. Furthermore, I propose that these rotations have their origin in the point couples of forces, denoted C, applied to this medium, where this medium is also not subject to any linear force. These point rotations are then transmitted point-to-point by the fact that each rotating medium point induces rotations in its neighboring points, and this results in the propagation of fields and particles. That these elastic medium points do not move linearly is due to the fact that, if there were such linear motion, then space would be linearly deformed. For example, drawing near to what is called the "Schwarzschild horizon", the spatial contraction would be infinite and, in this case, there would be no order in space. As such, it follows that **the only changes that the points of the medium can undergo are rotations of linearly immobile points in a medium that is called the "ether".** I denote $Par(m, e)$ any particle of mass, m, and electric charge, e.

The general lines of the unification demonstrated here include the following. The fact that the only changes that points in the ether can undergo are rotations of linearly-immobile points permits us to define, contrary to some other opinions, the fundamental notions of the **dimensions**, **time**, **velocity**, and **motion**. Thus, we may unify the fundamental notions of physics as changes in the ether. For example, time can be defined as proportional to the distance travelled by light in the free ether.

As such, I prove that electromagnetism is the case in which the ether is subject only to the densities $\boldsymbol{C}$ of couples of forces that create the field $\boldsymbol{\xi}$, associated with $Par(0,0)s$, i.e. to photons, of the rotations of the points of the ether and from which one can deduce the Maxwell equations and electromagnetic forces.

Electromagnetism is then generalized to the case where $\boldsymbol{\xi}$ is associated with $Par(m, e)s$, subject to incident fields by the fact that the Lagrange-Einstein function L_G of such a $Par(m, e)$ yields not only the equation of motion, but also ϕ_P, defined by $\hbar d\phi_P/dt = L_G$, which is the phase of a wave $\boldsymbol{\xi}$, associated with $Par(m, e)s$ and also ϕ related to ϕ_P, they are defined by

$$\phi = -t + \int d\ell/V$$

and

$$\phi_P = \omega(-t + \int d\ell/V_P)$$

where V denotes the group velocity, that is, the particle velocity and V_P is the phase velocity, as defined precisely in the following.

The wave $\boldsymbol{\xi}$ is a solution of a generalized ANSDE. A specific sum $R(\phi_P, \phi)$ of waves $\boldsymbol{\xi}$ forms "rotating ether trajectory segments" that move like $Par(m, e)$, containing all of its parameters. In fact, $Par(m, e)$ is $R(\phi_P, \phi)$. It appears that the fields are also elastic changes in the ether.

In the Newton-Maxwell theory, electrical repulsion between two massive electrically charged particles of the same sign is always greater than their Newtonian attraction. However, by taking into account general relativity and the ether theory, I show that their mutual gravitational attraction *surpasses* their electrical repulsion when they are *sufficiently close, one to the other*. This phenomenon plays the role of "strong nuclear interaction".

The Schrodinger equation ensues from axioms inspired from the de Broglie plane wave. As such, this equation is axiomatic, *even though it yields very important results.* Therefore, one may think that quantum mechanics may be generalized by a theory based on physical parameters and on general relativity, not just on axioms. Indeed, I show, in particular, that **Schrodinger's equation is a particular case of the ether elasticity** theory that is compatible with the general relativity, in particular the quantum states are shown to be due to interferences of the waves $\boldsymbol{\xi}$ that compose a $R(\phi_P, \phi)$ in an atom, i.e., where $R(\phi_P, \phi)$ describes a closed trajectory.

In linear motion, a $Par(m, e)$, i.e. a $R(\phi_P, \phi)$, creates a Lienard-Wiechert covariant potential tensor that yields the magnetic field $\boldsymbol{H}$, which, in *the ether elasticity theory, is the rotation velocity* $\partial_t \boldsymbol{\xi}$ *of the ether points*. It appears that, at a fixed observation point and at a given instant near to the moving electron, i.e., to the moving $R(\phi', \phi)$, the velocity in the ether is of the same form as the velocity of a point of a rotating solid, where the axis of such a rotation is parallel to the electron's trajectory. This phenomenon is electron spin, which, as I show, in the quantum state of an atom can take only quantized values.

Einstein's tensor $g_{\mu\nu}$ is usually considered to only represent the gravitational field. However, the Einstein equations that define $g_{\mu\nu}$ are pure mathematical reasoning, related only to covariant derivatives. Therefore, one can suppose that this tensor is more general in nature than only defining the gravitational field and thus can define other fields also, such as the EM field. I show that the fact that the electrostatic potential of force $A_{4,S}$ (S = static), created by an immobile electric charge e_0, is of the same form as the Newtonian potential of force $G_{4,S}$, created by an immobile $Par(m, e)$, which implies that $A_{4,S}$ is a particular $G_{4,S}$. This is generalized by the fact that the electromagnetic Lienard-Wiechert potential tensor A_μ, created at a point by a *moving* $\boldsymbol{e}$, is of the same form as the Newtonian approximation (NA) G_μ of $g_{\mu\nu}$, created by a *moving* m.

I generalize the Schwarzschild horizon radius by considering a spherical immobile mass m_0 of radius R_0 with a center located at a fixed ether point $\boldsymbol{O}$, and consider a particle of mass m_1, of velocity $\boldsymbol{V_1}$ due to the interaction of the fields created by m_0 and by m_1.

I prove how a mass is related to a field and what the constitution of that field is. The lines of this proof are based on the fact that, even when it is immobile, a mass m is related to a frequency ν_m by the relation $m = h\nu_m/c^2$. This shows that a mass m creates a field of frequency ν_m propagated in the ether. The nature of such a field is that it is an ensemble of points in the ether only rotating on themselves such that the axes of these rotations are situated on any line passing through the center of this mass.

I prove that on an axis joining two masses, m_1 and m_2, the fields of the rotation of the ether points due to m_1 and m_2 are of inverse senses. This means that they partially destroy themselves, i.e., the fields are smaller than those that are not between m_1 and m_2 and situated on this axis. It follows that m_1 and m_2 have the tendency to draw closer, that is, they are subjected to a force that tends to bring them closer together, finally forming only a single mass. This force is gravitational attraction.

Here, the "chemical" and "nuclear" reactions and the stability or instability of the atom are analyzed. In a chemical reaction involving several atoms, the number of atoms does not change, but they become connected, forming a unique molecule. Thus, a chemical reaction is an electronic reaction after such a reaction, the number of atoms remains the same, that is, all the nuclei involved in this reaction remain separated.

In contrast, a nuclear reaction, also occurring between several atoms, results in a "unique" atom such that <u>all</u> the nuclei of these atoms are fused into one nucleus. For example, one atom of helium H_e^2 results from two H atoms, the nucleus of this atom being constituted by the two H nuclei and neutrons. As such, in a nuclear reaction, protons are attracted to each other and form a new nucleus. More precisely, as has been shown to be due to the Schwarzschild attraction, in so-called "strong gravitation", two protons are attracted to each other when they are sufficiently close together, but repulse one another when their distance is greater than the threshold.

CHAPTER I

INTRODUCTION

I.1 On the Specific Elastic Medium Proposed by Maxwell, About Which Einstein Changed his Opinion and Finally Admitted its Existence, Calling it the "Ether"

In his treatise (Cf. Ref. 1, Art. 866), Maxwell explicitly presumed, although without developing the idea, the existence of a specific elastic medium in which electrodynamics is an ensemble of specific changes that express, in particular, the elastic nature of electrodynamics. Here is the relevant quotation (denoted Quot. 1).

> *Hence all these theories lead to the conception of a medium in which the propagation takes place, and if we admit this medium as a hypothesis, I think it ought to occupy a prominent place in our investigations, and that we ought to endeavor to construct a mental representation of all the details of its action, and this has been my constant aim in this treatise.*

In his 1905 paper (Cf. Ref. 2), Einstein refers to the *luminiferous aether* and **rejects the idea that such a medium can exist**. Indeed, here is the relevant quotation, denoted Quot. 2, with its first argumentation.

> *The introduction of a "luminiferous aether" (Lichtäther) will prove to be superfluous inasmuch as the view here to be developed will not require an "absolutely stationary space" provided with special properties, nor assign a velocity vector to a point of empty space in which electromagnetic processes take place.*

However, in 1920 Einstein completely changed his opinion about the possibility of the ether. Here is the relevant quotation, denoted Quot. 3, in its **second argumentation** as written in Ref. 3.

> *.... recapitulating, we may say that according to the general theory of relativity space is endowed with physical qualities; in this sense, therefore, there exists an ether. According to the general theory of relativity space without ether is*

unthinkable; for in such space there not only would be no propagation of light, but also no possibility of existence for standards of space and time (measuring-rods and clocks), nor therefore any space-time intervals in the physical sense.......

A very important deduction from Quot. 3 is that Einstein found that ***the existence of the ether permits us to precisely define the notions of time and space.*** *This fact is developed in Sec. I.2 and in Sec. II.1, below.*

In the past, numerous efforts were made to develop an elastic interpretation of physical phenomena and, in particular, of electrodynamic phenomena. Among these efforts, I can quote the works of Stokes, Navier, Cauchy, and Poisson, as well as many others. A review of these efforts is given by Sommerfeld (in Ref. 4; Cf. Ref. 5, pp. 106-113), in which an elastic model of the free Maxwell equations has been elaborated with only the electric field and the magnetic field being defined, but where the electric charges, the electric currents, the electromagnetic forces, and the electromagnetic energy have not been defined as elastic quantities. Furthermore, although a tentative elastic interpretation of Maxwell's equations of free electromagnetism was given, there was no attempt to give a tentative elastic interpretation to other physical phenomena. Such phenomena include, for example, the physical interpretation of particles; that of the fields applied to particles or created by them; that of the spin of the electron; that of the phenomena of quantum mechanics; and that of the strong nuclear interaction. Furthermore, there were no responses offered to some questions, including whether a relation exists between Maxwell's theory of electromagnetism and Einstein's theory of gravitation. This question can be asked since the electrostatic Coulomb field created by an immobile electrical charge is of the same form as Newton's approximation of the Schwarzschild field. Within the frame of these remarks and of this question, I show that these physical phenomena are due to the field of rotations, $\boldsymbol{\xi}$, of the point elements of an elastic medium, the ether, the points of which are always linearly immobile, but can rotate on themselves. Indeed, if there were ether points that were moving and had been translated, then, for example in the Schwarzschild case, there would be infinite translations of the ether points near the Schwarzschild horizon, as demonstrated in the following.

This confirms the fact that the ether points are linearly immobile, but can rotate on themselves. For example, an ether point that was at rest, begins to rotate due to an external cause and induces rotations on its linearly immobile neighboring points. These then rotate and induce rotations on their linearly immobile neighboring points, and so on.

This elastic medium cannot be other than the ether introduced by Maxwell and Einstein, as described in the above citations (quots. 1, 2, and 3) and confirmed above and in the following.

This field $\boldsymbol{\xi}$, defined above, is shown to be caused by densities of couples $\boldsymbol{C}$ of forces applied to the ether and also by other parameters. In particular, as I have shown in previous publications and do so again below, electromagnetism is related to $\boldsymbol{\xi}$, to $\boldsymbol{C}$, and to other parameters defined in the following, by the fact that $\boldsymbol{C}$ generates electrical charges and currents, also creating a specific field $\boldsymbol{\xi}$ from which one can deduce the electromagnetic field. We will see in the following that the elastic interpretation of electromagnetism gives a physical explanation of the electromagnetic action at distance and contributes to the completion of Maxwell's theory of electromagnetism and eliminates the inconsistency mentioned by Einstein in Ref. 6, of which I present below a relevant quotation, denoted Quot. 4.

> *The introduction of the field as an elementary concept gave rise to an inconsistency of the theory as a whole. Maxwell's theory, although adequately describing the behavior of electrically charged particles in their interaction with one another, does not explain the behavior of electrical densities, that is, it does not provide a theory of the particles themselves. They must therefore be treated as mass points on the basis of the old theory. The* combination of the idea of a continuous field with the conception of material points discontinuous in space appears inconsistent. *A consistent field theory requires continuity of all elements of the theory, not only in time, but also in space and in all points of space. Hence, the* material particle has no place as a fundamental concept in field theory. ***The particle can only appear as a limited region in space in which the field strength of the energy density is particularly high.*** *Thus, even apart from the fact that gravitation is not included, Maxwell's electrodynamics cannot be considered a complete theory.*

This elastic interpretation of electromagnetism constitutes a departure of the general elastic interpretation of other domains in physics and thus the ether elasticity theory, can be generalized to unify the diverse theories of physics by the fact that they will be shown to be particular cases of the ether elasticity theory.

Many physicists, particularly Einstein, have tried to unify the theories of physics, for example, seeking to unify electromagnetism and general relativity, but without success. In my opinion, this is due to the fact that these eminent physicists did not hypothesize any unique physical cause that connects these different domains of physics. That is to say, they did not see that electromagnetism, the constitution of particles, the constitution of fields, the nature and causes of electron spin, the nature and causes of the strong nuclear interaction, the nature of quantum mechanics, and the relation connecting general relativity and electromagnetism are all particular cases of a unique physical phenomenon. Therefore, I present here the unique physical phenomenon that connects these domains of physics.

In fact, these domains of physics are particular cases of ether point rotations. They are all particular states of ether point rotations, showing that these domains of physics are particular cases of the theory of ether elasticity. The ether is shown to be a specific elastic medium within which points can only rotate on themselves, but not move linearly, one relative to the other.

I present now the general and fundamental concepts of the ether, unifying the theories of physics. Among these concepts, **the theory of ether elasticity is an indispensable element**.

I.2 On the Concepts of Distance, Time, Mass, Phase Velocity, and Group Velocity in the Frame of the Ether

In free space, there is no vacuum, but there is a specific medium called the "ether". Numerous authors have predicted the existence of the ether, in particular, Maxwell and Einstein (Cf. the citations in quotations 1, 2, 3, and 4). This confirms the fact that the fundamental characteristic of space is that it is not a void, but constituted by the specific elastic medium called ether. The concept of the ether, as envisaged by Maxwell and Einstein (Cf. refs. 7-9), and developed further by the author (Cf. refs. 10-11), is presented here in detail. The ether appears to be an elastic medium within which its points can **only rotate elastically on themselves**, that is, points within the ether cannot move linearly one relative to the others. By the term "elastically", I mean that when the cause of this rotation no longer exists, then it disappears and this ether point returns to its free state. It appears that, contrary to some opinions, this concept of the ether permits us:

a) To define the fundamental notions of dimensions, time, and velocity.
b) To unify the fundamental notions of physics.

Indeed, the ether is a specific elastic isotropic medium with constant properties within which one can define the positions of points and the "distance" between any two fixed points. This important fact, namely where one can define the positions of points and their distances from each other, is analyzed below, since, in this case, the medium of the ether has to be induced with specific characteristics. I call **light** the effects of a perturbation created in a given domain of the ether and propagated out of this perturbed region. If the ether is not perturbed by another cause, then light always propagates in the same manner. The fact that the ether is an elastic isotropic medium with constant properties allows us firstly to define the fact that light always propagates in non-perturbed ether in the same manner and, as I go on to show, *specific rotational ether segments* can be formed in the ether.

In free space, light is always propagated in the same manner, i.e., propagated *independently from the motion of its emitter*, and this allows us to define the concepts of the ether and of other physical notions presented in the following. Firstly, I am going to define the notions of distance and of "time duration".

A unity δt of **time duration** can be defined as being **proportional to a given length** $\delta \ell$ travelled by a free photon in the ether, that is, by "free" light. For example:

> *One second will be proportional to the distance of value c, travelled by the photon in the free ether.*

As such, if for any arbitrary time duration Δt, during which a photon has travelled the distance $\Delta \ell$, which can be different from $\delta \ell$, then Δt and $\Delta \ell$ are related by the relation

$$\Delta t \equiv \Delta \ell / c \tag{1.1}$$

where c is a constant called free light velocity. Having considered the notion of time duration Δt and length $\Delta \ell$, we can define the notion of velocity, **temporarily**, and generally, denoted, **W**.

Definition: ***W*** *denotes the quotient of the length* ΔL *travelled through propagation and of time duration* Δt *during which this propagation was performed, that is*

$$W \equiv \Delta L / \Delta t \equiv c \Delta L / \Delta \ell \tag{1.2}$$

we can see, for example, that

$$[\Delta L \neq \Delta \ell\,] \Leftrightarrow\ [W \neq c] \tag{1.3}$$

However, since there are cases where the velocity $\boldsymbol{W}$ of a perturbation propagating in the free ether is different to c, we have to conceive of the existence of an entity called "mass". This mass causes such a perturbation to propagate at a velocity denoted $\boldsymbol{V}$, (instead of the general notation W), with an amplitude strictly smaller than c, even when it moves in the free ether. In fact, the unique reason why a free particle moves in the free ether at a velocity V strictly smaller than c, is that, this particle possesses a mass denoted generally m. Having conceived of the fact that the mass of a particle modifies its velocity, we now have to generalize and analyze the notion of velocity.

As shown in our previous publications and demonstrated here, in fact, **two different kinds of velocities** are associated with the **massive** particle, that is, $\boldsymbol{V}$ is associated with another velocity denoted $\boldsymbol{V_P}$. This is the reason why I have generally denoted velocity as W. However, we should note that for the massless particle, i.e., for the photon, these two kinds of velocity are identical.

These two different kinds of velocities associated with a massive particle are: the "group velocity", generally denoted $\boldsymbol{V}$; and the "phase velocity", generally denoted $\boldsymbol{V_P}$. As proven in our previous publications and below, V_P and V are not independent, but are connected by the relation

$$\frac{\partial}{\partial E_T}\frac{E_T}{V_p} = \frac{1}{V} \tag{1.4}$$

where E_T denotes the total particle energy defined by

$$E_T = mc^2 + h\nu \tag{1.5}$$

where ν denotes a frequency, and h, the Planck constant.

Let us consider now the case where the particle is a photon, that is, of zero mass, for which the velocity denoted V_{phot} does not depend on E_T. This fact implies that the left-hand member of Eq. (1.4) does not depend on E_T and therefore $V_{p.phot}$ does not depend on E_T, that is, for the photon, Eq. (1.4) becomes simply

$$V_{p.phot} = V_{phot} \tag{1.6}$$

In this case, where Eq. (1.6) is verified, then $V_{p.phot}$ and V_{phot} are generally denoted c_p (the index "p" in c_p is for "perturbed ether"), but are denoted simply c in the free ether. That is, for the photon, the phase and the group velocity are identical and denoted c_p or c. Furthermore, the expression for the photon energy denoted $E_{T,phot}$, considering (1.5) for $m = 0$, is

$$E_{T,phot} = h\nu \quad (1.7)$$

That is, by saying that a particle is different from a photon, I mean that this particle possesses a mass. As shown here, a photon is a particle of zero mass that moves at the velocity c_p or c, independently of its frequency ν, that is, independently of its energy $E_{T,phot}$. A massive particle moves at the group velocity V, which depends on its energy E_T. Another velocity, called the **phase velocity** and denoted V_P, is associated by the relation (1.4) with V, the function of which is defined in the following.

Definition

I call a "rotating ether trajectory segment" a segment of a trajectory in the ether that can be very small and where each point of this segment rotates around it, as demonstrated below. The ether points are linearly immobile and can only rotate on themselves, which is to say that only the locations of these rotations can change, but not the ether points. In particular:

> A "rotating ether trajectory segment" moving like light is a photon;
> a "free rotating ether trajectory segment" moving at a velocity smaller than that of the free photon, is a massive particle;
> and the photon is a massless particle.

As shown in our previous publications and developed further here, the ether is a particular elastic medium, the points of which **cannot move linearly** one relative to the other. The only change that a point of the ether can undergo is to rotate on itself. The rotation of a linearly immobile ether point induces rotations or rotational changes in its linearly immobile neighboring points, and these induced rotations or rotational changes induce rotations or rotational changes in other new linearly immobile ether points, and so on. *The propagation of such a field of **first rotations** is in fact the propagation of a particle.* Let us now consider the causes of such rotations, that is, what causes the **first rotations** of the ether points that influence their neighboring points?

> *It appears, as I prove, that these first rotations are fundamentally caused by point couples **C** of forces applied to some ether points that cause the rotations of their neighboring ether points, and so on.*

The rotations of the ether points are elastic, that is, when the causes of these rotations disappear, then they also disappear and the points of the ether return to their free state, that is, at a non-rotating state. As such, the ether is a medium in which each of its points can **only rotate elastically on itself**, but not move spatially relative to its neighbors.

A fundamental property of the ether is that these rotations are created by couples of forces denoted $\boldsymbol{C}$, applied to ether points, or by the ether points influencing other neighboring ether points (it is important not to confound couples of forces $\boldsymbol{C}$ with the light velocity c).

We may recall that, in the *general case*, an elastic medium is such that its points can be subjected to volumetric densities of linear forces $\boldsymbol{f}$ and to volumetric densities of couples of forces $\boldsymbol{C}$. These points in the medium may move linearly, one relative to the other, and also may rotate on themselves. The general equation of elasticity that governs these linear motions of its points and rotations on themselves, that is, the general elasticity equation that governs the field of the relative changes $\boldsymbol{\xi}$ of this general elastic medium ensues from the following **static** elasticity equation

$$\boldsymbol{curlC}/2 + (\sigma + \eta)\boldsymbol{grad}(div\boldsymbol{\xi}) + \eta\nabla^2\boldsymbol{\xi} + \boldsymbol{f} = 0 \tag{1.8}$$

This has to be written as a time-dependent equation, that is, the right-hand member has to contain time derivatives. I call Eq. (1.8) the "static Navier-Stokes-Durand equation of elasticity", in which σ and η denote the two so-called Lamé constants, and in which not only the densities of couples $\boldsymbol{C}$ applied to the elastic medium, but also the densities $\boldsymbol{f}$ of forces applied to it, are taken into account. E. Durand (Cf. Ref. 12) completed the Navier-Stokes equation by including the notion of densities of couples $\boldsymbol{C}$, which appears to be indispensable, i.e., this equation differs from the static Navier-Stokes equation by the term $\boldsymbol{curlC}$, which was not taken into account in the original Navier-Stokes equation of elasticity.

I show now that from Eq. (1.8), we can deduce the Maxwell equations. Indeed, since $\boldsymbol{f}$ is an **arbitrary** general density of linear forces applied to the elastic medium, we can choose the case where the expression for $\boldsymbol{f}$ is such that

$$(\sigma + 2\eta)\boldsymbol{grad}(div\boldsymbol{\xi}) + \boldsymbol{f} = 0 \tag{1.9}$$

In this case, the left-hand member of Eq. (1.8) becomes

$$\boldsymbol{curl}(\boldsymbol{C}/2 - \rho_0 c^2.\boldsymbol{curl\xi}) \tag{1.10}$$

Since this expression has the dimensions of a density of rotational acceleration that is of $\rho_0\partial_{tt}\boldsymbol{\xi}$, where

$$c^2 = \eta_0/\rho_0 \tag{1.11}$$

ρ_0 denotes the constant volumetric density of the ether, c, the light velocity, and where, therefore, η_0 is defined by $\eta_0 \equiv c^2\rho_0$, it follows that

$$\boldsymbol{curl}(\boldsymbol{C}/2 - \rho_0 c^2.\boldsymbol{curl\xi}) = \rho_0 \partial_{tt}\boldsymbol{\xi} \tag{1.12}$$

This equation is proved by the fact that, as demonstrated in our previous publications and in Sec. IV.1 below, from Eq. (1.12) we can deduce the Maxwell equations of electromagnetism. For example, by changing the variables

$$\boldsymbol{E} \equiv \eta_0 \boldsymbol{curl\xi} - \boldsymbol{C}/2 \text{ and } \boldsymbol{B} \equiv \rho_0 \partial_t \boldsymbol{\xi} \tag{1.13}$$

Eq. (1.12) becomes

$$\boldsymbol{curlE} + \partial_t \boldsymbol{B} = \boldsymbol{0} \tag{1.14}$$

Eq. (1.14) is one of Maxwell's equations of electromagnetism. Through further variable changes, from Eq. (1.12) we can also obtain **all** of Maxwell's equations.

Eq. (1.12) is generalized below to the general case in which the particle is massive and subject to any field of forces. Before this, let us consider the solution to Eq. (1.12). We may recall that a photon is a "rotating ether trajectory segment" moving at the velocity of light. I demonstrate in the following the expression for this "free photon rotating ether trajectory segment", denoted $\boldsymbol{R}(\phi_P, \phi)$

$$\boldsymbol{R}(\phi_P, \phi) \equiv \xi_0 \, SINC[(\Delta\omega/2)\phi] \cdot [\mathbf{e}_y \cos(\phi_P) + \mathbf{e}_z \sin(\phi_P)] \tag{1.15}$$

where, for the free photon, ϕ and ϕ_P are defined by

$$\phi \equiv -t + x/c, \quad \phi_P \equiv \omega(-t + x/c) \tag{1.16}$$

and where $SINC(\Psi)$ is defined by

$$SINC(\Psi) \equiv (sin\Psi)/\Psi \tag{1.17}$$

for any Ψ.

As such, the free photon $\boldsymbol{R}(\phi_P, \phi)$, as generalized in the following, can be written $\boldsymbol{R}(\omega\phi, \phi)$. We can see that $\boldsymbol{R}(\phi_P, \phi)$ is a function of x at a fixed instant t_0 and is a function of t at a fixed location x_0.

As demonstrated in our previous publications and here, I have adapted Eq. (1.8) to the time-dependent case in which the elastic medium is subject to point couples of forces $\boldsymbol{C}$. In this case, Eq. (1.8) becomes Eq. (1.12) and this elastic medium is called the ether. The reason why the linear forces $\boldsymbol{f}$ verify Eq. (1.9) is the fact that we can deduce the Maxwell equations from Eq. (1.12). Equation (1.12), from which we have obtained the Maxwell equations, is then generalized in the frame of the ether. I now present the fundamental ideas concerning this and mathematically develop them in the following.

We should note that for $\Delta\omega = 0$, that is, for

$$SINC[(\Delta\omega/2)\phi] = 1$$

Eq. (15) becomes

$$\boldsymbol{\xi} \equiv \xi_0 \left[\mathbf{e}_y \cos(\phi_P) + \mathbf{e}_z \sin(\phi_P)\right] \tag{1.18}$$

This equation represents a wave of ether point rotations from which we can deduce other electromagnetic quantities.

I.3 Two Examples: Group Velocity and Phase Velocity

Here we offer two examples: the first one deals with group velocity and the second one is for phase velocity. In the first example, I discuss the influence of the immobile particle on its environment. In the second, I consider the free massive or massless particle and its influence on its environment.

First example

We consider the case of a massive or massless particle subject to a field created by a point particle of mass m_0, immobile at the origin of the coordinates. This field, created by m_0, is a Schwarzschild field (Cf. Sec. V.8.2). We denote k as the constant of gravitation and a as the Schwarzschild constant, defined by

$$\alpha \equiv 2m_0 k/c^2 \tag{1.19}$$

where $\boldsymbol{r}$ is the radius vector; $\widehat{\mathbf{Vr}}$ is the angle made by $\boldsymbol{r}$ and the velocity $\boldsymbol{V}$ of a particle in this field created by m_0; and γ^2, $\hat{\gamma}^2$, B are the quantities defined by

$$\gamma^2 = 1 - \alpha/r, \hat{\gamma}^2 = 1 + \alpha(cos^2\widehat{\mathrm{Vr}})/(r\gamma^2) \qquad (1.20)$$

$$B = \sqrt{1 - (\gamma\, mc^2/E_T)^2} \qquad (1.21)$$

As demonstrated in our previous publications and below, the expressions for the particle velocity V and the phase velocity V_P of the wave associated with this particle in the field created by the immobile m_0 are

$$V = cB\gamma/\hat{\gamma} \qquad (1.22)$$

$$V_P = c\gamma/(\hat{\gamma}B) \qquad (1.23)$$

We can see that, for the photon, i.e., for $m = 0$, then the right-hand members of eqs. (1.22) and (1.23) are identical, since $B = 1$. As such, in this case V and V_P are identical, generally denoted c_p, for which the expression in this field is

$$c_p = c\gamma/\hat{\gamma} \qquad (1.24)$$

Having defined the phase velocity, we can deduce the density of the **phase** kinetic energy, denoted η, for which the expression is

$$\eta = \rho_0 {V_P}^2 \,. \qquad (1.25)$$

For example, the expression for the photon energy in a Schwarzschild field, which is simply its kinetic energy, since, for the photon, the phase and group velocities are identical, considering (1.24), is

$$\eta = \rho_0(c\gamma/\hat{\gamma})^2 \equiv \rho_0 {c_p}^2 \qquad (1.26)$$

In this case, Eq. (1.12) is partially generalized by the fact that c^2 has to be partially generalized by ${c_p}^2$, as defined in (1.26). Therefore, on the *Schwarzschild horizon*, i.e. for $r = \alpha$, then $c_P = 0$. It follows that the density of the **phase** kinetic energy η *is null*, and Eq. (1.12) becomes

$$\boldsymbol{curl}(\boldsymbol{C}/2) = -\rho_0\omega^2\boldsymbol{\xi} \equiv \rho_0\partial_{tt}\boldsymbol{\xi} \,. \qquad (1.27)$$

The question is thus: what happens when c_p is zero? Particularly considering that ρ_0 is a constant. The answer is determined by Eq. (1.26), which shows that when c_p is zero, then η is also zero, i.e., the rotations of the ether points do not move; this happens particularly on the Schwarzschild horizon.

Remark

This confirms the fact that ether points are linearly immobile and the movement is the **rotations** of the points. This is because if the ether points were moving and had translated their position, then, for example in the Schwarzschild case, there would be relative infinite translations of the ether points near the Schwarzschild horizon.

Second example

We consider now the case in which the massive or massless particle is free. Here, in eqs. (1.22) and (1.23) $\hat{\gamma}$ and γ take the value 1 and $m_0 = 0$, that is, these equations become

$$V = c\sqrt{1 - (mc^2/E_T)^2} \tag{1.28}$$

$$V_P = c/\sqrt{1 - (mc^2/E_T)^2)}\ . \tag{1.29}$$

We can see that a very small V, i.e., a very large mc^2 in front of $h\nu$, causes V_P to be very large, that V is always smaller or equal to c, and that $V = c$ implies $V_P = c$. The particle is then a free photon.

I.4 Preliminary Remarks on the Constitution of the Particle in the Frame of the Ether

As shown here, what I call a particle is, in fact, a rotating ether trajectory segment, generally denoted $\boldsymbol{R}(\phi_P, \phi)$, where ϕ_P denotes the phase of $\boldsymbol{\xi}$, as defined in the following

$$\phi_P = \frac{1}{\hbar}\left(-\int E_T dt + \int \frac{E_T}{V_P} d\ell\right)$$

and where ϕ is defined by

$$\phi = -t + \int d\ell/V \ .$$

$\boldsymbol{R}(\phi_P, \phi)$ is formed by the sum of waves of close phase velocities, of which, for example, the expression for the free photon is given in Eq. (1.15) and that for the **free massive** particle is also given by $\boldsymbol{R}(\phi_P, \phi)$. However, in the case of the free massive particle, the expressions for ϕ_P and ϕ are now given by

$$\phi = -t + x/V \qquad (1.30)$$

$$\phi_P = \omega(-t + x/V_P) \qquad (1.31)$$

which becomes Eq. (1.16) for the free photon.

This sum of phase velocities V_P moves at the group velocity V, which, in fact, is the particle velocity. This rotating ether trajectory segment $\boldsymbol{R}(\phi_P, \phi)$ moves at the group velocity V, propagating a wave called a "central phase wave", which propagates at the velocity V_P on this axis. The expression for $\boldsymbol{R}(\phi_P, \phi)$, which defines the particle, is defined by eqs. (1.15) and (1.16) for the free massless particle, i.e., a photon, and by the equations (1.15), (1.30), and (1.31) for the free massive particle. We can see that, in these cases, $\boldsymbol{R}(\phi_P, \phi)$ is a function of x at a fixed instant t_0 and is a function of t at a fixed point x_0, since, in these free cases, c, V, and V_P are constant. As demonstrated in the following, $\boldsymbol{R}(\phi_P, \phi)$ is the solution to an equation that generalizes Eq. (1.12) in a domain where $C = 0$. We can see that, for $m = 0$, **i.e., for the photon**, the two quantities ϕ_P/ω and ϕ, defined in eqs. (1.30) and (1.31), are identical. However, they are not identical for the massive particle.

I.5 The Elastic Nature of Mass and Gravitation

I.5.1 Generalities and introduction

Let us consider the case where, on the left-hand side of an ether point denoted P, there is an ether rotation, in one sense of rotation, and on the other side of P, i.e., on its right, there is the opposite rotation, but of similar amplitude. In this case, P, near to which are applied these two opposite rotations, remains immobile (cf. Fig. 1). However, if the left and right rotations are not of the same amplitude, P would not remain immobile. In this case, the question is: how does it happen that at the two sides of P, along the xaxis for example, the **amplitudes** of these ether point rotations are not equal? The answer to this question is that there is a supplementary effect that causes this non-equilibrium situation. This supplementary effect is because there is a supplementary point P_1, also with two rotations on the sides of the ether points such that between P and P_1 the total ether rotations are smaller than those that are outside this interval on the line that joins them. This fact, i.e., that between these two points the ether rotations are smaller, is due to the interference of the two rotations of inverse senses that exist between these two sources of rotation, which interfere and cause the forces that bring these two points closer together.

1.5.2 Generalization of the Schwarzschild case

We generalize the Schwarzschild radius $\alpha \equiv 2m_0k/c^2$ (Cf. Eq. (1.19)) by considering a spherical immobile mass m_0 of radius R_0 with its center located at a fixed ether point $\boldsymbol{O}$, and a particle of mass m, of velocity $\boldsymbol{V}$, due to the fact that it is subject to the field created by m_0. The expression for V is then (Cf. Eq. (1.22))

$$V = cB\gamma/\hat{\gamma} \tag{1.32}$$

where

$$\gamma = \sqrt{1 - R_0/r} \tag{1.33}$$

$$\hat{\gamma} = \sqrt{1 + R_0 cos^2(\widehat{\mathrm{Vr}})/(r\gamma^2)}\ . \tag{1.34}$$

In these expressions, $\boldsymbol{r}$ denotes the radius vector originating from the center of m_0 and directed towards the center of m; $\widehat{\mathbf{Vr}}$ denotes the angle made by $\boldsymbol{r}$ and $\boldsymbol{V}$; and B and Ω denote the quantities defined by

$$\Omega = \frac{mc^2}{mc^2 + h\Delta\nu} \tag{1.35}$$

$$B = \sqrt{1 - (1 - R_0/r)\Omega^2}\ . \tag{1.36}$$

In Eq. (1.35), c denotes the free light velocity and ν a frequency. It follows that Eq. (1.32) can then be written as

$$V = \frac{\sqrt{1-R_0/r}}{\sqrt{1+R_0 cos^2(\widehat{\nabla r})/(r\gamma^2)}}\sqrt{1-\left(1-\frac{R_0}{r}\right)\Omega^2}\,. \tag{1.37}$$

For $m = 0$, i.e., for $\Omega = 0$, Eq. (1.37) gives the photon velocity, denoted V_{ph}, defined by

$$V_{ph} = c\frac{\sqrt{1-R_0/r}}{\sqrt{1+R_0 cos^2(\widehat{\nabla r})/(r\gamma^2)}}\,. \tag{1.38}$$

In the free case, that is for $m_0 = 0$ and with $R_0 = 0$, V becomes V_{free}, defined by

$$V_{free} = c\sqrt{1-\Omega^2} \tag{1.39}$$

which, for $m = 0$, i.e., $\Omega = 0$, takes the value c.

1.5.3 The elastic nature of mass

Eq. (1.35) shows that mc^2 has the dimensions $h\nu$, and therefore we have the frequency ν_m, such that

$$mc^2 \equiv h\nu_m \tag{1.40}$$

and so Ω can be written

$$\Omega = \frac{\nu_m}{\nu_m+\Delta\nu}\,. \tag{1.41}$$

Considering eqs. (1.35), (1.39), and (1.41), it appears that:

> a free massive particle of mass m moves at velocity V_{free}, defined in (1.39), (1.35), and (1.41), which depends on the frequency ν_m and causes $|V_{free}| < c$; while a free massless particle, that is, a free photon, moves at velocity c independently of its frequency.

The explanation for these facts is that the mass modifies the medium in such a way that, for example, a free particle moves at velocity V_{free}, which is different to c. Since a free photon moves independently of its frequency, and V_{free} is defined in eqs. (1.39) and (1.41), it follows that, for the photon,

$\nu_m = 0$, i.e., m = 0. Since a free photon always moves at velocity c, it follows that the free photon moves independently of its frequency.

We consider here the physical constitution of the free mass m that can be immobile, and then consider the constitution of the photon, which moves at velocity c when it is free, but can move at a velocity less than c and can even be immobile, for example, when it is located on the Schwarzschild horizon.

The question remains as to how a mass is related to a field and what is the constitution of this field. The answer is that, even for an immobile mass m, one has (Cf. Eq. (1.40)) $m = h\nu_m/c^2$. This shows that an immobile mass m creates a field of frequency ν_m propagated in the ether. I term this phenomenon

mass wave of frequency $\nu_m = mc^2/h$

(Cf. refs.10, 13 and 14).

This mass wave concept is generalized to electrical charges and fields as specific changes in the medium of the ether such that they influence one another. Regarding, in particular, the field $h\nu_0/c^2$, created by an immobile mass m_0, we find that it **generalizes** the case treated by Einstein and Schwarzschild. Then, conceptually speaking in relation to the first ascertainment of the nature of the mass, we have

$$R_0 \equiv K\nu_0 \tag{1.42}$$

where K is a constant of dimensions $2hk/c^4$, that is,

$$2m_0k/c^2 = 2hk\nu_0/c^4 \implies m_0c^2 = h\nu_0 \tag{1.43}$$

As such, γ^2 and $\hat{\gamma}^2$ defined in (1.33) and (1.34) can be written as being the following functions of ν_0

$$\gamma^2 = 1 - K\nu_0/r \tag{1.44}$$
$$\hat{\gamma}^2 = 1 + K\nu_0(cos^2\widehat{\mathrm{Vr}})/(r - K\nu_0)$$

We can see that, for $\widehat{\mathrm{Vr}} = 0$, i.e., for $\boldsymbol{V}$ directed towards $\boldsymbol{O}$ or its opposite sense, we have

$$\hat{\gamma}^2 = r/(r - K\nu_0)\ . \tag{1.45}$$

Now, considering the expressions for γ^2 given in (1.44) and that for $\hat{\gamma}^2$ given in (1.45), we have

$$\gamma^2/\hat{\gamma}^2 = \frac{(r-K\nu_0)^2}{r^2} . \tag{1.46}$$

Thus, in the vector form along the x axis, where $\lfloor x \rfloor$ is the distance from the origin **O** to the point $\pm x$, and considering that $1 - K\nu_0/\lfloor x \rfloor \equiv \lfloor \Delta x \rfloor / \lfloor x \rfloor$, Eq. (1.37) becomes

$$\boldsymbol{V}(x) = \boldsymbol{e}_x c \frac{x}{\lfloor x \rfloor}\left(1 - \frac{K\nu_0}{\lfloor x \rfloor}\right)\sqrt{1 - \left(1 - \frac{K\nu_0}{\lfloor x \rfloor}\right)\Omega^2} . \tag{1.47}$$

For the photon, i.e., for $\Omega = 0$, Eq. (1.47) becomes

$$\boldsymbol{V}_{ph}(x) = \boldsymbol{e}_x c \frac{x}{\lfloor x \rfloor}\left(1 - \frac{K\nu_0}{\lfloor x \rfloor}\right) . \tag{1.48}$$

Equations (1.47) and (1.48) take the particular values

$$\boldsymbol{V}(K\nu_0) = \boldsymbol{V}_{ph}(K\nu_0) = 0 \tag{1.49}$$

$$\boldsymbol{V}(\infty) = \boldsymbol{e}_x c\sqrt{1 - \Omega^2} \tag{1.50}$$

$$\boldsymbol{V}_{ph}(\infty) = \boldsymbol{e}_x c . \tag{1.51}$$

We consider now the case where the particle of mass m is immobile, i.e., where $\Omega^2 = 1$, that is, where the particle has null kinetic energy. Then, (1.47) becomes

$$\boldsymbol{V}(x, \Omega = 1) = \boldsymbol{e}_x c \frac{x}{\lfloor x \rfloor}\left(1 - \frac{K\nu_0}{\lfloor x \rfloor}\right)\sqrt{\frac{K\nu_0}{\lfloor x \rfloor}} . \tag{1.52}$$

It follows that

$$\boldsymbol{V}(K\nu_0, \Omega = 1) = \boldsymbol{0}, \qquad \boldsymbol{V}(\infty, \Omega = 1) = \boldsymbol{0} . \tag{1.53}$$

As such, a particle of null kinetic energy in the field created by the immobile mass m_0 is immobile only in the following two cases: when it is located at $\lfloor x \rfloor = K\nu_0$; and when it is located at $\lfloor x \rfloor = \infty$.

Up to now, we have considered the expression for the velocity of a particle in the frame of the ether, where this velocity is that of a massive or massless particle free or under the influence of an immobile mass and in particular in a Schwarzschild field. Here, I am going to show that the concept of mass is of a completely different nature to that of the electric charge. Indeed, I am going to analyze and demonstrate the **physical effect** that causes two electric charges of different signs to attract one another and then the **physical effect** that causes two electrical charges of the same sign to repulse one another.

Now, before analyzing how **two masses** attract one another, I am going to consider the effects that an isolated immobile mass of radius R has on the ether.

I.5.4 The elastic natures of one isolated immobile mass, of two interactive masses, and of the gravitational forces

An immobile isolated mass creates a field of rotations of ether points such that the axes of these rotations are situated on any line passing through the center of the mass. These rotations do not depend on the angle of such a line relative to a specific one passing through the center of this mass. Figure 1 shows such a line passing through the center of a mass; the arrows indicate the axes and the senses of the vector rotations of the ether points along the line.

m

←←←←← ⊙ →→→→→

Fig. 1

Here, we consider that, at a given instant t, two masses m_1 and m_2, are disposed as in Fig. 2.

m_1 m_2

←←←← ⊙ →→→→ ←←←← ⊙ →→→→

Fig. 2

At instant t, the center of a spherical mass m_1 of radius R_1 is located at point X_1 on the x axis, and the center of mass m_2 of radius R_2 is located at point X_2, also on the x axis, such that, at this instant, the distance of $L > 0$ is defined by

$$L = X_2 - R_2 - (X_1 + R_1)\ . \qquad (1.54)$$

In Fig. 2, the arrows indicate the axes and also the senses of rotations of the ether points along a straight line passing through the two centers of mass.

We can see that, between m_1 and m_2 on the x axis, the fields of rotations of the ether points are of inverse senses, that is, they destroy themselves there and the fields are smaller than those that are not between m_1 and m_2, but situated on their extensions.

As a result, the masses m_1 and m_2 have the tendency to draw closer and finally come to be so close that they form only one mass. This explains gravitational attraction. Of course, if these moving masses do not move the first towards the second in a straight-line trajectory, they would move one around the other with not too great velocities. However, if at least one of them has a sufficiently large velocity, then they will deviate from their initial trajectories.

Let us consider what happens at point P_1 on the x axis, the position of which is defined by

$$P_1 = X_1 + R_1 + \Delta x \qquad (1.55)$$

where Δx is a variable such that

$$0 \leq \Delta x \leq L \qquad (1.56)$$

and what happens at point y is defined by

$$\Delta x + y = L \quad . \qquad (1.57)$$

In particular, let us consider the photon velocity $\boldsymbol{V}_{ph1}$ due to a mass m_1 located at point X_1 on the x axis such that $\boldsymbol{V}_{ph1}$ is directed towards $\boldsymbol{O}$**, or in the opposite sense.** Considering $\Omega \equiv 0$, Eq. (1.45) deals with the component along the x axis, where $\lfloor x \rfloor$ is the distance from X_1 to the point $X_1 +\ R_1 + \Delta x$ on the x axis

$$\lfloor x \rfloor = R_1 + \Delta x, \qquad R_1 =\ K v_1,$$

$$V_{ph1}[X_1 \pm (R_1 + \Delta x)] = \pm c \frac{R_1 + \Delta x}{\lfloor R_1 + \Delta x \rfloor}\left(1 - \frac{K v_1}{\lfloor K v_1 + \Delta x \rfloor}\right) \qquad (1.58)$$

and, in particular

$$V_{ph1}(X_1 \pm R_1) = 0 \quad (1.59)$$

$$V_{ph1}[X_1 \pm (R_1 + \infty)] = \pm c \quad (1.60)$$

denoting

$$V_{ph1}(X_1 + R_1 + \Delta x) \equiv W_1(\Delta x)$$

and

$$V_{ph2}(X_2 - L - R_2 + \Delta x) \equiv W_2(\Delta x)\,.$$

Then, from Eq. (1.58), and considering Eq. (1.42) and $R_1 \equiv K\nu_1$, we have

$$W_1(\Delta x) \ = \frac{c\Delta x}{K\nu_1+\Delta x} \equiv \frac{c\Delta x}{R_1+\Delta x} \quad (1.61)$$

and also considering that the radius R_2 of the mass m_2 is $R_2 \equiv K\nu_2$

$$W_2(\Delta x) = \frac{c(L-\Delta x)}{L+K\nu_2-\Delta x} = \frac{c(L-\Delta x)}{L+R_2-\Delta x} \quad (1.62)$$

Therefore

$$W_1(\Delta x) - W_2(\Delta x) = \frac{c\Delta x}{R_1+\Delta x} - \frac{c(L-\Delta x)}{R_2+L-\Delta x}\,. \quad (1.63)$$

In particular

$$W_1(0) - W_2(0) = -\frac{cL}{L+R_2} \quad (1.64)$$

$$W_1(\mathrm{L}) - W_2(\mathrm{L}) = \frac{cL}{R_1+L} \quad (1.65)$$

remembering that $R_1 \equiv K\nu_1$ and $R_2 \equiv K\nu_2$. Equations (1.64) and (1.65) follow from the facts that, in these cases, m_1 and m_2 attract each other. Furthermore

$$[L = 0,\ ie., \Delta x = 0] \Rightarrow W_1(0) - W_2(0) = 0\,. \quad (1.66)$$

In this last case, the two masses m_1 and m_2 are unified as one mass, $m = m_1 + m_2$, creating a new gravitational field.

We can see that there is no symmetry around the masses m_1 and m_2. More precisely, we can see that the ether rotations on the two sides of each of these masses are not symmetrical. This causes the masses m_1 and m_2 to move one towards the other until their surfaces touch each other.

Now, I seek to determine the mathematical relation between the fact that, for example, a symmetry or dissymmetry can occur around a spherical mass, according to the rotations of the ether points located on the two sides of a mass denoted as m. This is like a mass that would be pushed on its two sides such that it remains immobile, but if it were to be pushed by two non-exactly inverse forces, the mass would move. Therefore, I show that the rotation of the ether points near the mass is equivalent to a force applied to this mass. This is due to the fact that the only change that occurs on the ether points is that they can rotate, but not move. It is the rotations that are moving and therefore the gravitational field is constituted by the rotations of the ether points due to this field. As such, it is gravitation that creates rotations of ether points such that when they are immobile, the rotations are symmetrical around the mass.

I.5.5 The elastic nature of an isolated immobile electric charge versus that of two electric charges

Contrary to a mass, an electric charge can be positive or negative. I denote them e_+ and e_-. An electric charge is a small length in the ether such that each of its points has rotated, and these rotations are such that their axes are situated along this small length, as shown in the two following pictures

e_+

←←←←←←←←←←

Fig. 3

and

e_-

→→→→→→→→→→

Fig. 4

In these pictures, the arrows indicate the direction and senses of rotation of the ether points. It is evident that these rotations engender the rotation of ether points in their vicinity and that these rotations engender new ether point rotations, and so on. We can see that electric charges of the same sign repulse one another because the fields generated by these charges are rotations of the same sense and the sum of these rotations is greater than any one element of this sum. However, we can see that two electric charges of different signs will attract one another because the fields generated by these charges are rotations of inverse senses and the sum of these rotations is smaller than any one element of this sum.

I.5.6 Conclusion to Sec. I.5

Thus, the ether theory allows us to explain (Cf. Ref. 15) the gravitational attraction that occurs between any two masses, the attraction between two opposite electric charges, and the repulsion that occurs between two non-opposite electric charges.

I.6 Some Notations

Greek indices take the values 1,2,3,4, while Latin indices take the values 1,2,3. These last refer to spatial quantities, while the index value 4 refers to temporal quantities. I denote x^μ as the contra-variant coordinates; the light velocity in a vacuum by c; the temporal coordinates by x^4, i.e., $x^4 = ct$; and by ∂_μ, the partial derivative $\partial_\mu \equiv \partial/\partial x^\mu$, as well as $\partial_{t,} \equiv \partial/\partial t$. The rotation vector of a point of the elastic medium is denoted $\boldsymbol{\xi}$. A contra-variant tensor is denoted with the index up, e.g., A^μ, and a covariant one, with the index down, e.g., A_μ. The Einstein summation is used here not only with Greek indices, but also with Latin indices. As usual, $g_{\mu\nu}$ is the co-covariant of Einstein's fundamental tensor and ds the Einsteinian infinitesimal element, defined by

$$ds \equiv \sqrt{g_{\mu\nu} dx^\mu dx^\nu} \equiv \sqrt{dx^\mu dx_\mu}\ . \qquad (1.67)$$

Since ds, dx^μ, and $dt = dx^4/c$ are well defined, the ratios ds/dt and dx^μ/dt are also well defined. They will be denoted $\dot{s}, \dot{x}^\mu$, and more generally df/dt by $\dot{f}$. I denote $d\ell$ as the infinitesimal element of a ray of a wave, or that of a particle trajectory, using the notation

$$u^\mu \equiv dx^\mu/d\ell\ . \qquad (1.68)$$

By $\mathfrak{R}(g, A)$, I denote the spatial region in which an incident gravitational field $g_{\mu\nu}$ and an incident electromagnetic field A_μ can be present, but not the sources of these fields. In the absence of gravitation, i.e., in $\mathfrak{R}(g, A)$, g takes the value 0, while, in the absence of the electromagnetic field, A takes the value 0. Furthermore, $\mathfrak{R}(g, A)$ is denoted $\mathfrak{R}(g, A)_0$ when the incident fields are time independent. In this case, as shown further, the total particle energy denoted E_T is constant, $A_i = 0$, and one can choose the coordinates to be such that $g_{4j} = 0$. A particle of inertial rest mass m and of electric charge e will be denoted $Par(m, e)$; m and e may be zero in the case where the particle is a photon, or e can be zero, but not m in the case where the particle is massive and neutral. The plural of $Par(m, e)$ is denoted $Par(m, e)s$.

CHAPTER II

THE CONSISTENCY OF PRELIMINARY NOTIONS ABOUT THE ETHER WITH THE CONCEPTS OF TIME, VELOCITY, MASS, AND SPECIAL RELATIVITY

II.1 Preliminary Notions Concerning a Specific Elastic Medium: The Ether

The concept of the ether, envisaged in particular by Maxwell and Einstein (Cf. refs. 7-9), has been partially developed in refs. 10-11. In the current chapter, I show, firstly, that, contrary to some opinions, that when it is free, the ether is an elastic isotropic medium with constant properties in which light always propagates at the same "velocity" independently of the emitter's motion. The notion of the velocity of propagation in the ether is precisely defined below.

It is this constancy of the velocity of the propagation of light in the free ether that allows us to define the notion of time duration. Indeed, a **unity of time duration** can be defined as being proportional to a given length travelled by a photon in the free ether. For example, one second can be defined as being proportional to the distance of the numerical value c travelled by the photon in the free ether. The velocity of the photon, which is the same as the velocity of light, is therefore also c, and this is constant in the free ether. One can therefore define the velocity $\boldsymbol{V}$ of a particle. Indeed, if this particle travels the distance L, while the photon travels the distance L_{ph}, then the expression for the amplitude of the velocity V of this particle will be

$$V = cL/L_{ph}$$

It appears that special relativity relies on the hypothesis that the ether exists. In the case of the photon, which is a massless particle, the group and phase velocities are identical. However, these two kinds of velocity are different for massive particles. Furthermore, it appears that continuing Einstein's special relativity theory allows us to demonstrate the results of the Morley-Michelson experiment. In this experiment, it was found that there is *no change* to the density of the ether, but rather the distance *travelled by the photon* on the earth, which moves relative to the ether, changes.

Let us define more precisely the specific elastic medium called ether such that when it is free, i.e., if it is not subject to any perturbation, it is isotropic and has constant properties that are not dependent on location, nor on time. As shown in the following, the ether can be affected only by volumetric densities of couples of forces, denoted C. That is to say that the ether is a medium that is not linearly deformed, i.e., that each of its points can only rotate on itself, but not change its location in the ether. Since the ether is elastic, the effects of an external cause that perturbs it even a small amount, causing the rotation of some points in the ether, are propagated step by step. As shown in the following, the concept of the elastic medium called ether allows us to unify the fundamental theories of physics.

II.2 The Concepts of Time and Motion Defined in the Frame of the Ether

I show here that it is only in the frame of the ether that we can define the notion of time and of the propagation of perturbations. In this context, we may consider Einstein's opinion on this subject (Quot. 3). An important deduction from Einstein's opinion is that:

> **The existence of the ether permits to define the notions of time and of space.**

Indeed, since the ether is not linearly deformed, we can define the notion of length in it. This agrees with Einstein's opinion (Quot. 3). As such, if $\delta\ell$ denotes a given constant length in the ether and W a given constant parameter, as defined below, I term this the <u>unity of time duration</u> and denote it δt, the quantity being defined by

$$\delta t \equiv \delta\ell / W \tag{2.1}$$

If $\Delta\ell$ denotes a length that can be different to $\delta\ell$, travelled by some effect, then we can define the time duration and denote it Δt, which can be different to δt, with the number of W contained in $\Delta\ell$, that is

$$\Delta t = \Delta\ell / W \equiv \delta t \, \Delta\ell / \delta\ell \tag{2.2}$$

Since Δt denotes a duration of time, it can be denoted more explicitly as

$$\Delta t = t - t_0 \tag{2.3}$$

$\Delta\ell$ can also be denoted more explicitly

$$\Delta\ell = \ell - \ell_0 \tag{2.4}$$

that is, from (2.2), (2.3), and (2.4), we have

$$t - t_0 = \delta t\,(\ell - \ell_0)/\delta\ell\,. \tag{2.5}$$

Having defined the time duration Δt, we can define the instant t by the following

$$t = \ell/W \equiv \delta t\,\ell/\delta\ell \tag{2.6}$$

where t denotes the number of W contained in the length ℓ. It appears that W defined in (2.1), (2.2), and in (2.6), i.e.,

$$W = \Delta\ell/\Delta t \equiv \delta\ell/\delta t \equiv \ell/t \tag{2.7}$$

denotes a velocity. More precisely, W denotes the velocity of the propagation of perturbations in the free ether. These perturbations are created even in a small volume of the ether by couples of forces applied to points in the ether, or by the effects of perturbations created by the perturbed points in their vicinity. These perturbations are rotations of linearly immobile points in the ether, but not linear motions of ether points. The propagation of such induced rotations, that is, when each point of the ether rotates following a given order in the free ether, is called free "light" propagation. As such, I term "free ether" the case where the ether is not perturbed, or is perturbed only by light propagations. In such a case the light's velocity is constant and denoted c, rather than W. Propagations in the ether are always in the form of a sinusoidal wave or the sum of such waves of different but close frequencies that can form $\boldsymbol{R}(\phi_P, \phi)$, called photons, which move at the group velocity, which in the case of light, is also c. It follows that:

> *The existence of time is equivalent to the existence of an elastic isotropic homogeneous medium, the "ether", indeed:*
>
> *The existence of time implies the existence of the ether, since, in the absence of an ether, waves would not always be propagated with the same*

velocity and, therefore, one could not define time as being proportional to a length propagated by such waves.

Reciprocally, the existence of the ether permits us to define time, since, if one could not define the time, for example, taken by a wave to travel a given distance, one could not define the ether, since a time duration is proportional to a length in the ether.

If the velocity of propagation in the free ether is the "free light" velocity c, then let us consider Einstein's principle P_2, stated as follows (Cf. pp. 41of Ref. 7):

> *Any ray of light moves in the stationary system of coordinates with the determined velocity* c*, whether the ray is emitted by a stationary or a moving system.*

In fact, this principle implies the existence of the ether as defined here. Indeed, in the absence of an elastic medium, waves could not be propagated, but only particles and other entities the velocities of which would not be the same and independent of those due to emitters and of other causes that could move in empty space. Numerous authors predicted the existence of the ether, in particular, Maxwell and Einstein. Maxwell wrote (Art. 866 of Ref. 1):

> *... hence, all these theories lead to the conception of a medium in which the propagation takes place, and if we admit this medium as a hypothesis,* ***I think it ought to occupy a prominent place in our investigations****.*

Einstein (Cf. Sec. I above) stated:

> *.... according to the general theory of relativity,* ***space without ether is unthinkable****...*

It appears again that the stationary system cited by Einstein, i.e., the ether, is an elastic medium since light is propagated in it at velocity c, independently of the velocity of its emitter. If there were no such a medium, then the velocity of an emitted photon would depend upon the cause of this emission.

II.3 Generalization of the Notions of Velocity, i.e., Group Velocity and Phase Velocity

Having defined the notion of time in this ether, we now define the velocity $V(m)$ of a particle of mass m. Such a particle moving at velocity $V(m)$ is denoted $Par[m, V(m)]$. By stating that $Par[m, V(m)]$ moves at velocity $V(m)$, I mean that during time duration Δt, during which the particle covers distance $\Delta\ell(m)$, light covers the distance $\Delta\ell(0)$, that is

$$\Delta t = \Delta\ell(0)/c = \Delta\ell(0)\delta t/\delta\ell \tag{2.8}$$

and thus it follows that

$$V(m) = \frac{\Delta\ell(m)}{\Delta t} = \frac{\Delta\ell(m)}{\Delta\ell(0)} \cdot \frac{\delta\ell}{\delta t} = \frac{\Delta\ell(m)}{\Delta\ell(0)} c\ . \tag{2.9}$$

We can see, in particular, that $V(0) = c$. However, as shown in our previous publications and in Sec. V.4 below, in the ether, in addition to the particle velocity $V(m)$, there is another velocity called "phase velocity", denoted $V_P(m)$, related to $V(m)$ by (1.4). It can be rewritten

$$\frac{\partial}{\partial E_T}\left(\frac{E_T}{V_P(m)}\right) = \frac{1}{V(m)} \tag{1.4}$$

where E_T denotes the total particle energy and such that $V_P(0) = V(0) = c$, i.e., they are equal for the photon. However, for $m \neq 0$, the particle velocity $V(m)$, which is also the group velocity (Cf. Sec. V), is different to the phase velocity $V_P(m)$. We can see that, since $V_P(0) = c$, the velocity of the free photon does not depend upon its energy.

II.4 The Morley-Michelson Experiment

In the Morley-Michelson (MM) experiment, two photons with different trajectories meet (Cf. eqs. (36)-(40) in Ref. 11). Following the perpendicular trajectory, the time $t_\perp$ taken by the photon to return to its point of departure on the earth, that one repeats moves relatively to the ether, is

$$t_\perp = 2L/\sqrt{c^2 - V^2} \tag{2.10}$$

and, in parallel to the earth's motion, the time t_x taken by the photon to return to its point of departure on the earth, is

$$t_x = 2cL_x/(c^2 - V^2) . \tag{2.11}$$

Since these two photons meet, then

$$t_\perp = t_x \tag{2.12}$$

that is

$$L_x = L\sqrt{1 - V^2/c^2} . \tag{2.13}$$

Using notation $L(0)$ instead of L, and $L(V)$ instead of L_x (Cf. Eq. (32) of Ref. 11), we have

$$L(V) = L(0)\sqrt{1 - V^2/c^2} . \tag{2.14}$$

II.5 The Results of the Morley-Michelson Experiment Explained by Special Relativity

Let $\boldsymbol{e}_i$ (i=1,2,3) denote an orthonormal system K, considered as immobile in the ether, and $P(\boldsymbol{V})$ a point moving within it at velocity $\boldsymbol{V}$, i.e., immobile on the earth at velocity $\boldsymbol{V}$ relative to the ether. An observer located at $P(\boldsymbol{V})$ will see that a photon moving in the same direction as $\boldsymbol{V}$ has the velocity $\boldsymbol{V}_{\boldsymbol{ph+}}$, defined by

$$\boldsymbol{V}_{\boldsymbol{ph+}} = \boldsymbol{c} - \boldsymbol{V} \tag{2.15}$$

while a photon moving in the opposite direction will be seen by this observer to have the velocity $\boldsymbol{V}_{\boldsymbol{ph-}}$, defined by

$$\boldsymbol{V}_{\boldsymbol{ph-}} = -\boldsymbol{c} - \boldsymbol{V} . \tag{2.16}$$

On the earth, we may consider a photon located at $P(\boldsymbol{V})$ at time t_0 and moving in the same direction as $\boldsymbol{V}$. After having covered distance x' on the earth, it is reflected and returns to its departure point $P(\boldsymbol{V})$ on the earth. Let τ denote the time duration measured by a watch fixed at $P(\boldsymbol{V})$, i.e., fixed on the earth, taken by the photon to cover distance x' in the direction of $\boldsymbol{V}$ and then returning to its point of departure $P(\boldsymbol{V})$, for which the distance is also x' on the earth. As defined by Einstein (pp. 44, Ref. 7), duration τ is in particular a function of $\frac{x'}{c-V}$ and of $\frac{x'}{c+V}$. Now let us consider more profoundly the physical significance of these quantities. In this context, I use the following relations

$$\frac{x\prime}{c-V} = \frac{x\prime + V\Delta t_1}{c} = \Delta t_1 \tag{2.17}$$

$$\frac{x\prime}{c+V} = \frac{x\prime - V\Delta t_2}{c} = \Delta t_2 \tag{2.18}$$

where Δt_1 respectively Δt_2 denotes the time, measured by a clock that is immobile in K, that the photon takes to travel distance x' measured on the earth in the direction of its motion respectively in the opposite direction. We can verify that

$$\frac{x\prime}{c-V} - \frac{x\prime + V\Delta t_1}{c} = \frac{V}{c}\left(\frac{x\prime}{c-V} - \Delta t_1\right) \tag{2.19}$$

having the equivalence

$$\frac{x\prime}{c-V} = \Delta t_1 \Leftrightarrow \quad \frac{x\prime}{c-V} = \frac{x\prime + V\Delta t_1}{c} \tag{2.20}$$

and, in the same way, having the following equivalence

$$\frac{x\prime}{c+V} = \Delta t_2 \Leftrightarrow \frac{x\prime}{c+V} = \frac{x' - V\Delta t_2}{c}\,. \tag{2.21}$$

These terms, $\frac{x\prime}{c-V}$ and $\frac{x\prime}{c+V}$, appear in Einstein's equation (pp. 44, Ref. 7), namely

$$\tau(0, t_0) + \tau\left(0, t_0 + \frac{x'}{c-V} + \frac{x'}{c+V}\right) = 2\tau\left(x', t_0 + \frac{x'}{c-V}\right) . \tag{2.22}$$

As written by Einstein, x' is very small and so this equation can be written as

$$\frac{\partial \tau}{\partial x'} + \frac{V}{(c^2 - V^2)}\frac{\partial \tau}{\partial t} = 0 \tag{2.22.1}$$

(Cf. the third equation, pp. 44, Ref. 7). As we can verify, the solution of Eq. (2.22.1) is

$$\tau = \sqrt{1 - (V/c)^2}\left(t - \frac{x'V}{c^2 - V^2}\right) . \tag{2.23}$$

In order to obtain Eq. (2.23), I repeated Einstein's reasoning (pp. 44, Ref. 7), using the first equation found on pp. 45, there.

As we can see, x' is the constant length on earth travelled by a photon in one direction and then in the opposite direction, that is, the photon has travelled $2x'$. This length x' is constant on earth with velocity V, and it is the constancy of x' that is **fundamental** to this line of reasoning. Indeed, t, as it appears in Eq. (2.23) is a parameter that needs to be defined so that we can properly define τ. I am going to define t in such a way that Eq. (2.23) becomes identical to Eq. (2.10), i.e., we shall determine t such that (2.23) becomes

$$\tau = 2x'/\sqrt{c^2 - V^2} \tag{2.24}$$

which is of the same form as Eq. (2.10), where τ is no other than $t_\perp$ and x' is no other than L. In order to determine the expression of t, such that (2.23) becomes (2.24), we have to determine t as a solution of the equation

$$\sqrt{1 - (V/c)^2}\left(t - \frac{x'V}{c^2 - V^2}\right) = \frac{2x'}{\sqrt{c^2 - V^2}}\,. \tag{2.25}$$

The solution of this equation, being the value of t such that (2.23) becomes (2.24), is

$$t = x'\,(2c + V)/(c^2 - V^2)\,. \tag{2.26}$$

Thus, by inserting the expression for t given in (2.26) into Eq. (2.23), we can obtain Eq. (2.24), which is identical to the result (2.10) of the MM experiment where $\tau \equiv t_\perp$ and $x' \equiv L$. It follows that:

Special relativity explains the results of the Morley-Michelson experiment.

II.6 Interpretation of the Results of the Morley-Michelson Experiment in the Frame of the Ether Theory

It follows from the MM experimental results, that the distance $L(0)$ travelled by the photon when this length is immobile relative to the ether becomes $L(V)$, as defined by (2.14), when this length is on the earth, moving at velocity V relative to the ether in which the photon moves at velocity c. Looking at Eq. (2.24), which is the same as eqs. (2.10) and (2.11), we can obtain Eq. (2.14), defining the change of the length of the trajectory travelled by the photon on earth, moving at velocity V relative to the ether. The question remains as to what represents this change of length from $L(0)$ to $L(V)$ in Eq. (2.14). In other words: the length that changes, is the length of what?

In order to respond to this question, we should note that, as demonstrated above, the photon travels distance $L(V)$ defined in (2.14) on the earth, moving at velocity V relative to the ether. It appears that there are two eventual responses to the question regarding the change of length from $L(0)$ to $L(V)$. These two eventualities are as follows:

Eventuality 1: the length $L(0)$ on the earth is physically modified and becomes $L(V)$, i.e., it becomes the length travelled by the photon on the earth.

Eventuality 2: the length $L(0)$ on the earth remains physically unchanged, but the photon travels the distance $L(V)$ on the earth.

In order to answer this question and show what represents the length $L(V)$ on the earth, let us consider the particular case where, in these two eventualities, the earth moves at the velocity of light c, i.e., $V = c$.

In Eventuality 1, the distance $L(V)$ becomes $L(c) = 0$ (Cf. Eq. (2.14)). That is to say that, on the one hand, the length on the earth becomes zero, but contains the photon as well as all the passengers that are immobile on the earth. On the other hand, these passengers see that they have the same dimensions as they had when they were immobile relative to the ether, since this hypothetical change in length can only be seen by observers that are immobile in the ether, i.e., not on the earth.

In Eventuality 2, the photon also travels 0 distance on the earth, i.e., remains immobile there, but the length on the earth seen by an observer immobile in the ether, or seen by an observer immobile on the earth, remains unchanged at its value $L(0)$, ***since every point on the earth moves at the same velocity***. Indeed, since every point on the earth, in the frame of this approximation, moves at the same velocity, there can be no contraction in length because if there were, certain points on the earth would move at a velocity that was different to c.

It appears, therefore, that Eventuality 1 is unrealistic since a length is seen, at the same time, to be zero by an immobile observer in the ether and to have the length $L(0)$ by a passenger on earth. In contrast, Eventuality 2 appears realistic. Indeed, it is obvious that, if the earth moves at velocity c, then the photon remains immobile on the earth and a passenger immobile on the earth or immobile in the ether, sees that the length on the earth remains $L(0)$, since ***every point on the earth moves at the same velocity c**, supposed for this reasoning.* Finally, it appears that:

> *$L(V)$ defined in (2.14), denotes the **length travelled by the photon on the earth** that moves at velocity V relative to the ether while the length on the earth remains unchanged at the value $L(0)$, which it had when it was immobile relative to the ether.*

From this fact, an important conclusion follows that confirms our preceding discussion concerning the ether without dark matter:

> *It appears that, as shown in my previous publications, the ether's density remains unchanged.*

II.7 Conclusion to Sec. II

Contrary to what has been written by some authors, I demonstrate that we can define time univocally, for example, as being proportional to the distance travelled by light divided by a known constant parameter, such as c, the velocity of light. Since light is in the form of waves propagated in free space always at the same velocity, there must exist an "elastic" medium in which these waves are propagated. If there were no such medium, then there would be no waves propagated there. Even if the waves were there, they would not be propagated at the same velocity independently of the emitter. The existence of such an elastic isotropic homogenous medium, i.e., the ether, was predicted by Maxwell, Einstein, and a number of other physicists, and has been developed further by myself.

It is the notions of the ether and of the propagation of waves in it, i.e., of light, that permits us to define time and, reciprocally, it is the notion of time that allows us to define the ether. Indeed, time can be defined as being proportional to the distance of the propagation of vibrations in the ether and, reciprocally, distance can be defined as being proportional to the time taken by a photon to cover this distance.

I have presented an enlargement of the four-coordinate transformation from a stationary system to another system in a uniform translation relative to the former. If an observer on earth were to measure the distance between the point of departure of a photon and, for example, a mirror at a fixed location there, they would find this distance to be $L(0)$. If this observer, being on earth and immobile relative to the ether, were to send out a photon, he would find that the distance travelled by the photon was also $L(0)$.

However, if the earth were to move at velocity V relative to the ether and the photon at velocity c relative to the ether, an immobile observer on earth would find that the distance travelled by this photon would not be $L(0)$, but rather $L(V) = L(0)\sqrt{1-(V/c)^2}$. We should, however, note that $L(V)$ *is not a new dimension on the earth, but the distance travelled by a photon on the earth.* Thus, if the earth were to move at velocity c relative to the ether, the photon would be seen to be immobile by a passenger that was immobile on the earth, since $L(c) = 0$. It follows that, in the absence of dark matter, *the density of the ether remains unchanged.*

CHAPTER III

ELASTICITY THEORY: THE ETHER AS A PARTICULAR ELASTIC MEDIUM

An elastic medium with all of its points are in their rest states can be said to be a free elastic medium. The case in which a field of changes occurs in an elastic medium means that at each influence of perturbed points of the elastic medium acting on their neighbors. The external causes that can be applied together or separately to points of an elastic medium are of two kinds:

a) An ensemble of point couples of forces denoted $\boldsymbol{C}$.
b) An ensemble of point single force denoted $\boldsymbol{f}$

The general equation that governs the field of **static** changes $\boldsymbol{\xi}$ in an elastic isotropic and homogeneous medium of inertial density ρ, subjected to a time-independent field of point couples of forces $\boldsymbol{C}$, and to a time-independent field of single forces $\boldsymbol{f}$ (Cf. Ref. 12) is presented in Eq. (1.8). This can be rewritten as

$$\boldsymbol{curlC}/2 + (\sigma + \eta)\boldsymbol{grad}(div\boldsymbol{\xi}) + \eta\nabla^2\boldsymbol{\xi} + \boldsymbol{f} = 0 \qquad (3.1)$$

where σ and η denote the two Lamé coefficients. I call this very important equation the "Navier-Stokes-Durand **static** equation of elasticity". We can see that this equation not only takes into account the densities $\boldsymbol{f}$ of single forces applied to points of the elastic medium, but also the densities $\boldsymbol{C}$ of couples of forces applied to the same or to different points of this elastic medium. It was E. Durand at Université Paul Sabatier, Toulouse, France, who (Cf. Ref. 12) completed the Navier-Stokes equation by introducing the **indispensable** notion of **densities of couples of forces $\boldsymbol{C}$** applied to the elastic medium. As we can see, Eq. (3.1) differs from the Navier-Stokes equation by the term $\boldsymbol{curlC}/2$, which was not taken into account in the original Navier-Stokes equation of elasticity.

In the **non-static case**, i.e., in the **time-dependent case**, equation (3.1) is generalized by the fact that the right-hand member of this equation is not zero, but is a function of the time derivatives of $\boldsymbol{\xi}$ and also of other parameters defined in our previous publications and noted below.

I am going to consider some particular time-dependent cases here. Since a particular time-dependent case ensues from the general case of elasticity, it follows that there is also an ensemble of changes in the elastic specific medium called the **ether**, as defined in the following.

What we call the ether is a conservative elastic medium that can be defined by the fact that there is neither annihilation nor creation of what it is constituted. This fact is expressed by the following equation

$$div(\rho\partial_t\boldsymbol{\xi}) + \partial_t\rho = 0 \tag{3.2}$$

called the "continuity equation of the medium".

From the fundamental equation (3.1), we can deduce the Maxwell equations and, more generally, the equations that govern physical phenomena. Indeed, since $\boldsymbol{f}$ **is an arbitrary general density of forces** applied to the elastic medium, we can choose the case in which the expression for $\boldsymbol{f}$ is presented in Eq. (1.9). This can be rewritten as

$$(\sigma + 2\eta)\boldsymbol{grad}(div\boldsymbol{\xi}) + \boldsymbol{f} = 0\ . \tag{3.3}$$

As such, the ether is an elastic medium governed, in addition to an equation in which the left-hand member is defined by the left-hand member of Eq. (3.1), modified by Eq. (3.3). Thus, the left-hand member of Eq. (3.1) becomes

$$\boldsymbol{curlC}/2 + \eta[\nabla^2\boldsymbol{\xi} - \boldsymbol{grad}(div\boldsymbol{\xi})]$$

and since

$$-\nabla^2\boldsymbol{\xi} + \boldsymbol{grad}(div\boldsymbol{\xi}) \equiv \boldsymbol{curl}(\boldsymbol{curl\xi})$$

this left-hand member becomes

$$\boldsymbol{curlC}/2 - \eta\boldsymbol{curlcurl\xi} \tag{3.4}$$

However, in the **non-static case,** the right-hand member of Eq. (3.1), i.e., of (3.4), is not zero. Equation (3.4) shows that the ether is not subjected to deforming forces $\boldsymbol{f}$, as defined by Eq. (3.3). It follows that the density of inertia ρ is **constant** and this is then more precisely denoted ρ_0.

In the presence of an electromagnetic field, due to the creators $\boldsymbol{C}$ of this field, Eq. (3.4) becomes

$$\boldsymbol{curl}[\boldsymbol{curl\xi} - \boldsymbol{C}/(2\eta_0)] = -(\rho_0/\eta_0)\partial_{tt}\boldsymbol{\xi}\,. \tag{3.5}$$

This is due to the fact that, as discussed further in Sec. IV, from Eq. (3.5) we can deduce the Maxwell equations of electromagnetism, that is, Eq. (3.5) unifies these Maxwell equations. The general equation that governs the **ether** in the time-dependent case, i.e., in the case where the right-hand member of Eq. (3.1) is not zero, and in a domain where $\boldsymbol{C}$ is absent, is

$$[\nabla^2\boldsymbol{\xi} - \boldsymbol{grad}(div\boldsymbol{\xi})] = \left[\frac{\rho_0}{\eta}\partial_{tt} - \partial_\ell\left(\sqrt{\frac{\rho_0}{\eta}}\right)\partial_t\right]\boldsymbol{\xi} \tag{3.6}$$

where ∂_ℓ denotes the derivative along the particle trajectory. Eq. (3.6) can be written

$$-\boldsymbol{curl}(\boldsymbol{curl\xi}) = \left[\frac{\rho_0}{\eta}\partial_{tt} - \partial_\ell\left(\sqrt{\frac{\rho_0}{\eta}}\right)\partial_t\right]\boldsymbol{\xi}\,. \tag{3.7}$$

Considering further (1.25) and the monochromatic case, i.e., where

$$\partial_t\boldsymbol{\xi} = -i\omega\boldsymbol{\xi}$$

Eq. (3.6) becomes

$$\boldsymbol{curl}(\boldsymbol{curl\xi}) = \left[\frac{\omega^2}{V_P{}^2} - i\omega\left(\partial_\ell\frac{1}{V_P}\right)\right]\boldsymbol{\xi} \tag{3.8}$$

and thus Eq. (3.8) can be written as

$$\boldsymbol{curl}(\boldsymbol{curl\xi}) = \frac{\omega}{V_P{}^2}(\omega + i\partial_\ell V_P)\boldsymbol{\xi}\,. \tag{3.9}$$

The expression for ρ_0/η, which is the inverse of the square of the phase velocity V_P, will be analyzed below. We can state that in the free ether or if subjected to pure electromagnetic fields, η is the constant η_0 the value of which is defined by the relation (1.11).

CHAPTER IV

THE ELASTIC INTERPRETATION OF ELECTROMAGNETISM

IV.1 The Elastic Interpretation of the Maxwell Equations

We consider now the case where the ether is such that η is constant, thus being denoted η_0, i.e., the case where the ether is governed by Eq. (3.5). It appears that η_0 is the elasticity-restoring rotation coefficient, which is constant in the electromagnetic case. In the following, we will see that there are cases where the elasticity-restoring rotation coefficient is not constant. These will be denoted more generally by η, but ρ_0 is always a constant.

Equation (3.5) is the "electromagnetic ether elasticity equation". It defines the field $\boldsymbol{\xi}$ of the rotations of the ether points created by the density $\boldsymbol{C}$ of couples of forces applied to the ether.

I demonstrate now the following proposition.

Proposition

The case of elasticity where the ether is governed by Eq. (3.5) yields the Maxwell equations of electromagnetism.

Demonstration

Let us consider the following changes of variables

$$\boldsymbol{E} \equiv \eta_0 \boldsymbol{curl\xi} - \boldsymbol{C}/2 \qquad (4.1)$$

$$\boldsymbol{H} \equiv \partial_t \boldsymbol{\xi} \qquad (4.2)$$

$$\boldsymbol{B} \equiv \rho_0 \partial_t \boldsymbol{\xi} \qquad (4.3)$$

$$\boldsymbol{j}_e \equiv \partial_t \boldsymbol{C}/(2\eta_0) \qquad (4.4)$$

$$\rho_e \equiv -div\boldsymbol{C}/(2\eta_0) \,. \tag{4.5}$$

Inserting (4.1) and (4.3) into Eq. (3.5), we obtain

$$\boldsymbol{curlE} + \partial_t \boldsymbol{B} = \boldsymbol{0} \,. \tag{4.6}$$

Partially differentiating Eq. (4.1) with respect to time, and taking (4.4) and (4.2) into account, we get

$$\boldsymbol{curlH} - \partial_t \boldsymbol{E}/\eta_0 = \boldsymbol{J}_e \tag{4.7}$$

and then from (4.1), we have

$$div\boldsymbol{E} \equiv -div\boldsymbol{C}/2 \,.$$

Considering, furthermore, (4.5) we obtain

$$\rho_e \equiv div\boldsymbol{E}/\eta_0 \,. \tag{4.8}$$

From eqs. (3.5) and (4.3), we can write

$$div\boldsymbol{B} = 0 \,. \tag{4.9}$$

The four equations, (4.6), (4.7), (4.8), and (4.9), are identical to the Maxwell equations, thus demonstrating the validity of the proposition.

Furthermore, from equations (3.2), (4.3), and (4.9), the following relation is confirmed

$$\partial_t \rho_0 = 0 \tag{4.10}$$

and from (4.4) and (4.5), we deduce that $\boldsymbol{J}_e$ and ρ_e verify, as expected, the continuity equation

$$div\boldsymbol{J}_e + \partial_t \rho_e = 0 \,. \tag{4.11}$$

Thus, we can see that electromagnetism is a particular case of elasticity defined by the relations (3.1), (3.3), and (3.5), where:

$\boldsymbol{E}$, defined in (4.1), classically called the electric field, denotes the volumetric density of couples of forces applied to the ether;

$1/\eta_0$, classically termed the coefficient of electric induction in a vacuum, defined above, is such that η_0 denotes the elasticity-restoring rotation coefficient;
$\boldsymbol{H}$, defined in (4.2), classically termed the magnetic field, denotes the speed of the rotation of $\boldsymbol{\xi}$ points;
$\boldsymbol{B}$, from (4.3), classically termed magnetic induction, is simply defined by $\rho_0\mathbf{H}$, where ρ_0 denotes the volumetric density of inertia of the ether;
$\boldsymbol{J}_e$, defined in (4.4), denotes the volumetric density of electric currents;
ρ_e, defined in (4.5), denotes the volumetric density of electric charges.

We can see that ρ_e and $\boldsymbol{J}_e$, defined in (4.5) and (4.4), satisfy the continuity equation (4.11) of electric charges and currents. The coefficients ρ_0 and η_0 are such that

$$\eta_0/\rho_0 = c^2$$

where c denotes the free light velocity. We can see that ρ_0 and c being defined implies that η_0 is defined by $\eta_0 \equiv \rho_0 c^2$, and that η_0 has the dimension of the density of energy.

This case of elasticity, which is identical to Maxwell's electromagnetism, will be called the "elastic interpretation of electromagnetism". I now extend the elastic interpretation to electromagnetic forces and energy.

IV.2 The Elastic Interpretation of the Magnetic Force

I am going to demonstrate that the magnetic force that acts on an electric charge results from the interaction of $\boldsymbol{B}$, defined in Eq. (4.3), i.e., proportional to the speed of the rotation field, and of the speed of the rotation field $\boldsymbol{H_e}$, defined in Eq. (4.2), created by this charge and denoted e.

According to (4.3) and (4.2), and denoting $\partial_t\boldsymbol{\xi}_e \equiv \boldsymbol{H}_e$, we can define the following three quantities

$$p_r \equiv \boldsymbol{B} \cdot \boldsymbol{H}_e \equiv \rho_0 \partial_t \boldsymbol{\xi} \cdot \partial_t \boldsymbol{\xi}_e \qquad (4.12)$$

$$\boldsymbol{f}_{pr} = -\iint_S \boldsymbol{n} p_r dS \qquad (4.13)$$

$$\boldsymbol{f}_{reac} = \frac{1}{2}\iint\limits_S dS[(\boldsymbol{nB})\boldsymbol{H}_e + (\boldsymbol{nH}_e)\boldsymbol{B}]$$

$$\equiv \frac{\rho_0}{2}\iint_S dS[(\boldsymbol{n}\partial_t\boldsymbol{\xi})\partial_t\boldsymbol{\xi}_e + (\boldsymbol{n}\partial_t\boldsymbol{\xi}_e)\partial_t\boldsymbol{\xi}] \qquad (4.14)$$

where $\boldsymbol{n}$ denotes the unity normal vector going out from a close surface denoted S, which surrounds the electric charge as closely as we want, the infinitesimal element of S being denoted dS. We can see that p_r has the dimensions of a pressure; that $\boldsymbol{f}_{pr}$ has those of a force resulting from the pressure p_r; and that $\boldsymbol{f}_{reac}$ has those of a force resulting from the reaction of a pressure on S. Now we come to the following proposition.

Proposition

The magnetic force that acts on an electric charge e, of velocity $\boldsymbol{V}$, in an incident magnetic induction $\boldsymbol{B}$, is the same as if two forces were acting on this electric charge: the pressure force $\boldsymbol{f}_{pr}$ defined in (4.13) and in (4.12), and the reaction force $\boldsymbol{f}_{reac}$ defined in (4.14).

Demonstration

The expression, in MKSA units, for $\boldsymbol{H}_e$ is (Cf. eqs. (63.8) & (63.9), Ref. 16)

$$\boldsymbol{H}_e \equiv \partial_t\boldsymbol{\xi}_e = \frac{e\eta\boldsymbol{R}}{4\pi c\rho_0 R(R-\boldsymbol{R}\cdot\boldsymbol{V}/c)^3}\wedge\left\{\frac{(\boldsymbol{R}-R\boldsymbol{V}/c)}{\beta^2}+\frac{\boldsymbol{R}\wedge[(\boldsymbol{R}-R\boldsymbol{V}/c)\wedge\dot{\boldsymbol{V}}]}{c^2}\right\} \qquad (4.15)$$

where $\beta^2 \equiv 1/(1-V^2/c^2)$; $\dot{\boldsymbol{V}}$ denotes the acceleration of the electric charge; and $\boldsymbol{R}$ is the radius vector going from the point where the charge is located to the point of observation. We may consider that the current point of observation $\boldsymbol{r}$ is a point of a surface S' that would be the surface S if the electric charge were at the origin of the spatial coordinates. Since S is very small, one may replace $\boldsymbol{R}$ by $\boldsymbol{r}$, and neglect, in the brackets of (4.15), the term that has $\dot{\boldsymbol{V}}$ as a factor versus the other term, because this last term varies like $1/r$, while the first varies like $1/r^2$. Furthermore, taking $\boldsymbol{R}\wedge\boldsymbol{R}=0$ into account, on S we have

$$\boldsymbol{H}_e \equiv \partial_t\boldsymbol{\xi}_e = \frac{e}{4\pi\beta^2}\frac{\boldsymbol{V}\wedge\boldsymbol{r}}{(r-\boldsymbol{r}\cdot\boldsymbol{V}/c)^3}\,. \qquad (4.16)$$

I choose the z axis to be such that, at the given instant, $\mathbf{V} = V\mathbf{e}_z$ and suppose that S is rotationally symmetrical around $\boldsymbol{V}$. As a result, we have

$$\boldsymbol{r} = (\boldsymbol{e}_x cos\varphi + \boldsymbol{e}_y sin\varphi)Q + \boldsymbol{e}_z Z \tag{4.17}$$

$$\boldsymbol{n} = \frac{1}{\sqrt{1+Q'^2}}(\boldsymbol{e}_x \text{cos} + \boldsymbol{e}_y \text{sin} + \boldsymbol{e}_z \text{Q'}) \tag{4.18}$$

$$dS \equiv dzd\varphi Q\sqrt{1+Q'^2} \tag{4.19}$$

In these equations, Q denotes the distance of a point of S to the z axis and is a function of z, $Q' \equiv dQ/dz$, and φ denotes the angle made by the two planes defined by $(\boldsymbol{e}_r, \boldsymbol{e}_z)$ and $(\boldsymbol{e}_x, \boldsymbol{e}_z)$. The incident magnetic induction $\boldsymbol{B}$ *is spatially constant in S since it is very small.* Taking into account Eq. (4.12), eqs. (4.16)-(4.19), and the fact that $\boldsymbol{B}$ *is spatially constant in S*, after an elementary calculation, their integration (4.13) gives

$$\boldsymbol{f}_{pr} = \frac{a'}{4\beta^2} e\boldsymbol{V} \wedge \boldsymbol{B} \equiv \frac{a'}{4\beta^2} e\boldsymbol{V} \wedge \rho_0 \partial_t \boldsymbol{\xi} \tag{4.20}$$

where

$$a' \equiv \int_{z_1}^{z_2} Q^2 dz/(\sqrt{Q^2+z^2} - zV/c) \; . \tag{4.21}$$

In the same manner, the integral (4.14) gives $\boldsymbol{f}_{reac} \equiv \boldsymbol{f}_{pr}$. Therefore, the magnetic force denoted $\boldsymbol{f}_{B,V}$, which acts on an electric charge e of velocity $\boldsymbol{V}$ in an incident magnetic induction $\boldsymbol{B}$, is

$$\boldsymbol{f}_{B,V} = \boldsymbol{2f}_{pr} \; . \tag{4.22}$$

The factor a', defined in (4.21), depends on S, which thus depends on the form of the particle that has the electric charge. When $a/(2\beta^2) = 1$, we come back to the Lorentz magnetic force. Here we remark that the force $\boldsymbol{f}_{B,V}$ given in (4.22) is the force that acts on a *point* electric charge e, and not the force that would act on an electric charge extended in a non-zero volume inside the surface S. This is because, in this case, the expression for the magnetic force is much more complicated than the one given in Eq. (4.22). However, there is a case in which the charge is not a point, but we shall not address it here. This case is the one where the electric charge is uniformly distributed on a segment of a straight line that moves with velocity $\boldsymbol{V}$ directed in the sense of this segment, which is also, for example, that of the z axis, and where, in addition, $V << c$. In this case, I wish to show that the expression for this force is simply $\boldsymbol{f}_{B,V} = e\boldsymbol{V} \wedge \rho_0 \partial_t \boldsymbol{\xi}$. In effect, it is known that the expression for the magnetic field, denoted by $\boldsymbol{H}_{seg}$, created by this segment in its near-vicinity, is

$$\boldsymbol{H}_{seg} = \frac{e\boldsymbol{V}}{2\pi Q} \wedge \frac{\boldsymbol{Q}}{Q} \tag{4.23}$$

where

$$\boldsymbol{Q} \equiv Q[(cos\varphi)\boldsymbol{e}_x + (sin\varphi)\boldsymbol{e}_y] \,.$$

Substituting $\boldsymbol{H}_e$ by $\boldsymbol{H}_{seg}$ in eqs. (4.12) and (4.14),

$$\boldsymbol{f}_{B,V} = e\boldsymbol{V} \wedge \boldsymbol{B} \,.$$

Furthermore, (4.23) shows that the magnetic field $\boldsymbol{H}_{seg}$ is, in the near vicinity of the segment, the same as the *flow of a vortex. It reacts with an incident magnetic induction, that is to say, with an incident flow of the ether points, in an analogous manner with that which occurs in fluid mechanics* (see Ref. 17, vol. II, pp. 25-35). Therefore, I can also say that an electric point particle creates in its near vicinity the same flow as *an infinitesimal element of a vortex in a fluid.*

IV.3 The Elastic Interpretation of the Coulomb Potential

Here, I demonstrate that the Coulomb potential of interacting forces between two point electric charges is the energy of the interaction of the changes due to the densities of couples associated with these charges when one of them is immobile. Let us consider two point electrical charges, denoted e_0 and e_1, where e_0 is immobile at the origin. Let us denote $\boldsymbol{E}_0$ and $\boldsymbol{E}_1$ the electric fields created by e_0 and e_0. The expression for $\boldsymbol{E}_0$ is

$$\boldsymbol{E}_0 = -\frac{\eta_0 e_0}{4\pi} \boldsymbol{grad}\left(\frac{1}{r}\right) \tag{4.24}$$

Let us consider the integral $U_{0,1}$, defined by

$$U_{0.1} \equiv \frac{1}{\eta_0} \iiint \boldsymbol{E}_0 \boldsymbol{E}_1 dV_{ol} \equiv \frac{e_0}{4\pi} \int_0^\infty \frac{dr}{r^2} \iint \boldsymbol{n} \boldsymbol{E}_1 dS \tag{4.25}$$

where the triple integral is taken on $B[0,r]$ and the double one on $S[0,r]$, respectively, the sphere and the spherical surface centered on **0** and of radius

r. I denote $u(x)$ the unity step function; $\delta_3(\boldsymbol{r})$ the tridimensional Dirac distribution; and $\boldsymbol{r}_1$ the location of the charge e_1. Since the flux of a vector is equal to the volume integral of its divergence, and, since $div\boldsymbol{E}_1 = e_1\eta_0\delta_3(\boldsymbol{r}-\boldsymbol{r}_1)$, we have

$$U_{0.1} \equiv \frac{e_0e_1\eta_0}{4\pi}\int_0^\infty \frac{u(r-r_1)}{r^2}dr = \frac{e_0e_1\eta_0}{4\pi r_1}\,. \qquad (4.26)$$

Equation (4.26) shows that $U_{0,1}$ is the Coulomb potential. Therefore, the Coulomb potential is the energy of interaction of the two fields, $\boldsymbol{E}_0$ and $\boldsymbol{E}_1$. That is to say, considering (4.1) and (4.5), the Coulomb potential is the interaction energy of changes arising from a couple associated with these electric charges.

IV.4 The Elastic Interpretation of Electromagnetic Energy

Let us consider N electric charges e_n (n = 1, 2,..., N). Each e_n creates an electric field denoted $\boldsymbol{E}_n$ and the magnetic fields denoted $\boldsymbol{H}_n$. The electromagnetic energy denoted E_{EM} due to these charges is the energy of N fields $\boldsymbol{E}_n$ and of N fields $\boldsymbol{H}_n$. I denote $\boldsymbol{H}_T \equiv \sum_{n=1}^{N}\boldsymbol{H}_n$ and $\boldsymbol{E}_T \equiv \sum_{n=1}^{N}\boldsymbol{E}_n$. The expression for E_{EM} is then

$$E_{EM} = \frac{1}{2}\iiint (\rho_0 {H_T}^2 + \varepsilon_0 {E_T}^2)dV_{ol}\,. \qquad (4.27)$$

Considering the relation (4.2), I denote $\boldsymbol{H}_T = \partial_t\xi_T$. Thus, we have

$$\frac{\rho_0}{2}\iiint {H_T}^2 dV_{ol} = \frac{\rho_0}{2}\iiint \partial_t{\xi_T}^2 dV_{ol}\,. \qquad (4.28)$$

Relation (4.28) shows that the magnetic part of the right-hand member of Eq. (4.27) is the kinetic energy of *ether* point rotations.

Let us consider the electrical part of the right-hand member of Eq. (4.27). It is well known that, since $\partial_t\boldsymbol{\xi}$ is a field of rotational velocities, then $(1/2)\boldsymbol{curl}\partial_t\boldsymbol{\xi}$ is a function of the field of rotational velocities. Considering relation (4.1), we can deduce that η is proportional to the elasticity-restoring rotation coefficient, since $\eta\boldsymbol{curl\xi}$ and $\boldsymbol{E}$ have the dimensions of a volumetric density of couples, also having the dimensions of a rotation multiplied by an elasticity-restoring rotation coefficient. It then follows that $(\eta_0/2){E_T}^2dV_{ol}$ is the energy due to the rotations of ether points in dV_{ol}.

Considering formula (4.27), the elastic interpretation of the electromagnetic energy is as follows.

The energy of an electromagnetic field is the sum of two energies:
1 - The kinetic energy of the linear changes of the ether points.
2 - The rotational energy of the ether points.

We know that, if the electric charges are point charges, then the integral (4.27) is infinite. However, as we will see, it is not E_{EM} that interests us, but its derivatives.

IV.5 The Elastic Interpretation of Electromagnetic Waves

We consider a spatial domain containing an electromagnetic field, but not the electric charges that create this field, that is to say, where $\boldsymbol{C} = 0$ in eqs. (4.4) and (4.5). We consider, furthermore, the case in which the electromagnetic field is a monochromatic wave of frequency ν. In this case, Eq. (3.5) becomes

$$\boldsymbol{curl}(\boldsymbol{curl\xi}) = (\omega^2/c^2)\boldsymbol{\xi} \qquad (4.29)$$

for which, as I demonstrate, the solutions take the form

$$\boldsymbol{\xi} = \boldsymbol{\xi}_0 \exp\,(i\phi_P) \qquad (4.30)$$

where $\boldsymbol{\xi}_0$ is a constant vector perpendicular to the x axis, i.e.,

$$\boldsymbol{\xi}_0 = (0\,,\ \xi_{0y}\,,\ \xi_{0z}) \qquad (4.31)$$

and where ϕ_P denotes the phase of the wave $\boldsymbol{\xi}$, for which the expression is

$$\phi_P = \omega(-t + x/c) \qquad (4.32)$$

demonstrating that $\boldsymbol{\xi} = \boldsymbol{\xi}_0 \exp\,[i\omega(-t + x/c)]$ is the solution to Eq. (4.29).

Indeed, denoting

$$f \equiv exp(i\omega x/c) \qquad (4.33)$$

$$\beth' \equiv (\omega^2 f/c^2)\boldsymbol{e}_2 \qquad (4.34)$$

eqs. (4.30) and (4.29) become

$$\boldsymbol{\xi} = \boldsymbol{\xi}_0 f exp(-i\omega t) \tag{4.35}$$

$$\boldsymbol{curl}(\boldsymbol{curl\xi}_0 f) = \beth'\xi_0 \,. \tag{4.35.1}$$

Since $\boldsymbol{\xi}_0$ is constant and follow the direction of $\boldsymbol{e}_2$, Eq. (4.35.1) becomes

$$\boldsymbol{curl}[(\boldsymbol{grad}f)\wedge\boldsymbol{e}_2] = \beth' \tag{4.36}$$

now, considering Eq. (4.33), and that

$$\boldsymbol{grad}f = i\omega f e_1/c \tag{4.36.1}$$

(4.36) becomes

$$i\omega\boldsymbol{curl}(f\boldsymbol{e}_1\wedge\boldsymbol{e}_2)/c = \beth'$$

that is,

$$i\omega\boldsymbol{curl}(f\boldsymbol{e}_3)/c = \beth' \,. \tag{4.37}$$

However,

$$\boldsymbol{curl}(f\boldsymbol{e}_3) \equiv f\boldsymbol{curle}_3 + [\boldsymbol{grad}(f)]\wedge\boldsymbol{e}_3$$

i.e.,

$$\boldsymbol{curl}(f\boldsymbol{e}_3) \equiv [\boldsymbol{grad}(f)]\wedge\boldsymbol{e}_3 \tag{4.38}$$

therefore, considering (4.38), (4.37) becomes

$$i\omega\boldsymbol{grad}(f)\wedge\boldsymbol{e}_3 = \beth' \,. \tag{4.39}$$

We should recollect that $\boldsymbol{grad}f = i\omega f e_1/c$, and, introducing this expression in (4.39), we obtain

$$f\omega^2\boldsymbol{e}_2 = c^2\beth' \,. \tag{4.40}$$

Q.E.D.

From (4.1), (4.2), and (4.30)-(4.32), we obtain

$$\boldsymbol{E} = i\eta_0(\omega/c)(0\,, -\xi_{0z}\,,\ \xi_{0y})exp(i\phi_P) \tag{4.41}$$

and

$$\boldsymbol{H} = -i\omega\boldsymbol{\xi}_0 \exp(i\phi_P)\ . \tag{4.42}$$

We can see that $\boldsymbol{E}$ is perpendicular to $\boldsymbol{H}$, and that $\boldsymbol{\xi}_0$, $\boldsymbol{E}$, and $\boldsymbol{H}$ are perpendicular to the x axis, i.e., to the light ray.

I show that the photon is constituted by a superposition of waves. I denote this by ϕ and call it the "group phase", the quantity being defined by

$$\phi \equiv \hbar\partial\phi_P/\partial E_T \tag{4.43}$$

and since for the photon $E_T = h\nu$, considering (4.32), it follows that

$$\phi = -t + x/c\ . \tag{4.44}$$

IV.6 On the Constitution of the Free Photon in the Frame of Ether Elasticity

In Sec. I.2, I introduced the notion of "rotating ether trajectory segments". Such a segment trajectory can be very small and each of its points rotates in such a way that the axes of these rotations are located on this trajectory. However, these ether points are linearly immobile and the movement is the self-rotation of the ether points, defined more completely below. That is to say, that only the locations of these rotations can change, but not of the ether points themselves. In particular:

> A free photon is a "free rotating ether trajectory segment" moving at free light velocity.
> A free massive particle is a "free rotating ether trajectory segment" moving at a velocity smaller than that of a free photon.

As shown in our previous publications and developed further in this one, the ether is a particular elastic medium. Its points **cannot move linearly**, one relative to the other; the only change that a point in the ether can undergo is to rotate on itself.

I show now that the **free** photon is a rotating ether trajectory segment, denoted $\boldsymbol{R}(\phi_P, \phi)$, which is a function of ϕ defined in (4.44), and of ϕ_P defined in (4.32), for which the expression is

$$\boldsymbol{R}(\phi_P, \phi) \equiv \xi_0\, SINC[(\Delta\omega/2)\phi] \cdot \left[\mathbf{e}_y \cos(\phi_P) + \mathbf{e}_z \sin(\phi_P)\right] \tag{4.45}$$

where $SINC(\Psi)$ is defined in Eq. (1.17). We can see that $\boldsymbol{R}(\phi_P, \phi)$ is a function of x at a fixed instant t_0 and a function of t at a fixed point x_0. Let us now consider the component of this rotating ether trajectory segment $\boldsymbol{R}(\phi_P, \phi)$ in the plane (x, z), that is, let us consider the expression for $\boldsymbol{R}(\phi_P, \phi) \cdot \mathbf{e}_z$. Considering (4.45), we have

$$\boldsymbol{R}(\phi_P, \phi) \cdot \mathbf{e}_z \equiv \xi_0\, SINC(\Delta\omega\phi/2) \cdot \sin(\phi_P) \tag{4.46}$$

$$[\boldsymbol{R}(\phi_P, \phi)\mathbf{e}_z]\mathbf{e}_z = \mathbf{e}_z\xi_0\, SINC(\Delta\omega\phi/2) \cdot \sin(\phi_P)\,. \tag{4.47}$$

I show now that $[\boldsymbol{R}(\phi_P, \phi)\mathbf{e}_z]\mathbf{e}_z$ is a specific sum of waves. To this end, we can use the following notations. When $\boldsymbol{\xi}$, ϕ_P, and ϕ, which are functions of fixed ω, become functions of $\omega + \Delta\omega$, they are denoted, specifically, $\boldsymbol{\xi}(\Delta\omega)$, $\phi_P(\Delta\omega)$, and $\phi(\Delta\omega)$. However, following the finite increment theorem, for the photon, we can write

$$\phi_P(\Delta\omega) = \phi_P + \Delta\omega\phi \tag{4.48}$$

i.e.,

$$\boldsymbol{\xi}(\Delta\omega) = \boldsymbol{\xi} exp(i\Delta\omega\phi)\,. \tag{4.49}$$

Let us consider the following superposition $\hat{\boldsymbol{\xi}}(\Delta\omega)$ of waves $\boldsymbol{\xi}(\vartheta)$

$$\hat{\boldsymbol{\xi}}(\Delta\omega) \equiv \frac{\mathbf{1}}{\Delta\omega}\int_{-\Delta\omega/2}^{\Delta\omega/2} \boldsymbol{\xi}(\vartheta)d\vartheta\,. \tag{4.50}$$

By virtue of (4.49), Eq. (4.50) becomes

$$\hat{\boldsymbol{\xi}}(\Delta\omega) \equiv \frac{\mathbf{1}}{\Delta\omega}\boldsymbol{\xi}\int_{-\Delta\omega/2}^{\Delta\omega/2} exp(i\vartheta\phi)d\vartheta \tag{4.50.1}$$

where ϕ is defined in Eq. (4.44). This is to say that

$$\hat{\boldsymbol{\xi}}(\Delta\omega) \equiv \boldsymbol{\xi}_0\, SINC[(\Delta\omega/2)\phi]\exp\,(i\phi_P) \tag{4.51}$$

and where ϕ_P is defined in (4.32). Equation (4.51) can be written as

$$\hat{\boldsymbol{\xi}}(\Delta\omega) \equiv \boldsymbol{\xi}_0 SINC\left[\left(\frac{\Delta\omega}{2}\right)\phi\right](cos\phi_P + isin\phi_P)$$

which is also

$$-i\hat{\boldsymbol{\xi}}(\Delta\omega) \equiv [\boldsymbol{R}(\phi_P, \phi)\mathbf{e}_z]\mathbf{e}_z - i\boldsymbol{\xi}_0 SINC[(\Delta\omega/2)\phi]cos\phi_P \tag{4.52}$$

We can see that

$$Real\ part\ of\ \left[-i\hat{\xi}(\Delta\omega)\right] = [\boldsymbol{R}(\phi_P,\phi)\boldsymbol{e}_z]\boldsymbol{e}_z \qquad (4.53)$$

for which the expression is given in Eq. (4.47). $[\boldsymbol{R}(\phi_P,\phi)\boldsymbol{e}_z]\boldsymbol{e}_z$ is the expression for the component in the fixed plane (x, z) of $\boldsymbol{R}(\phi_P,\phi)$ defined in Eq. (4.45). That is to say, Eq. (4.45) generalizes Eq. (4.51) and represents a free massless particle, being a free photon, in which c denotes the free light velocity, i.e., the free photon velocity. Such a particle, i.e., a photon, is also symmetrical around the x axis.

At time t and point x on the x axis, the expression for the rotation $\boldsymbol{R}(\phi_P,\phi)$ of this photon is given in Eq. (4.45). We can see that the maximal value of this rotation (ϕ_P,ϕ) is that of $\boldsymbol{R}(0,0)$, defined by

$$\boldsymbol{R}(0,0) \equiv \xi_0\ \mathbf{e}_y \qquad (4.54)$$

for which its absolute value is

$$|\boldsymbol{R}(0,0)| \equiv \xi_0\ . \qquad (4.55)$$

We define the free photon's length and its rotational distribution. These rotations depend on their emplacement in the rotating ether trajectory segment. The free photon length is defined from the point on the axis where the rotation is at its maximum, i.e., from the point where

$$SINC\left[\frac{\Delta\omega}{2}\left(-t+\frac{x}{c}\right)\right] = 1 \qquad (4.56)$$

to the point on the axis where the rotation is very small. This is where

$$SINC\left[\frac{\Delta\omega}{2}\left(-t+\frac{x}{c}\right)\right] \text{ is very small}$$

i.e., where it takes a very small value ϵ when x takes the value x_ϵ, defined by

$$SIN\left[\frac{\Delta\omega}{2}\left(-t+\frac{tc+x_\epsilon}{c}\right)\right] \cong 2c/(\Delta\omega x_\epsilon) \cong \epsilon\ . \qquad (4.57)$$

The expression for the length denoted x_ϵ of this rotating ether trajectory segment of the free photon is defined by

$$x_\epsilon \cong 2c/(\Delta\omega\epsilon)$$

where x_ϵ denotes the distance going from the point in the photon where the rotation is at its maximum, i.e., from the point $|\boldsymbol{R}(0,0)|$, to the point in the photon where the rotation is very small. This is where the free particle rotating ether trajectory segment is $|\boldsymbol{R}(\phi_P, \phi)|$, defined by

$$|\boldsymbol{R}(\phi_P, \phi)| \cong \epsilon\xi_0$$

$\boldsymbol{R}(\phi_P, \phi)$, *defined in* (4.45), *and called a single-photon wave, represents a photon for which the amplitude is non zero only in the volume defined by* $SINC\left[\frac{\Delta\omega}{2}\left(-t+\frac{x}{c}\right)\right]$ *that moves at velocity c. Thus, the single-photon wave* $\boldsymbol{R}(\phi_P, \phi)$ *describes all the physical properties of the photon, i.e., its trajectory, its velocity, its energy, its undulating mechanical nature, and its size.*

As shown above, a photon is the sum of electromagnetic waves and behaves as a small hard projectile, described as the **photo-electric effect**.

Notation

To avoid confusion, the changes $\boldsymbol{\xi}$ defined by equations (3.5) and (4.30), i.e., associated with photons, i.e., with $Par(0,0)$, are also denoted $\boldsymbol{\xi}_{0,0}$.

IV.7 Interpretation of the Effects of the Photon Rotating Ether Trajectory Segment on its Vicinity

I showed above that a photon is a "rotating ether trajectory segment" $\boldsymbol{R}(\phi_P, \phi)$ of limited length, in which the ether points rotate and the rotational axes follow the trajectory with moving m at velocity V where m can be zero. The interesting aspect is that these rotating points induce rotations in their neighbors, which then induce rotations in their neighbors, and on into their neighbors, etc.

These induced rotations are in fact the field created by this rotating ether trajectory segment.

We now consider the case in which the field is created by $\boldsymbol{R}(\phi_P, \phi)$ of zero mass and is also of null electrical charge. This field is constituted by rotating ether points in which the rotational axes of the ether points are aligned in the same direction as the trajectory of the rotating ether trajectory segment. The relevant ensemble of such rotations creates a field of the rotation of the ether around the trajectory of this charge. This field is circular when the electron trajectory is a straight line and is simply composed by closed trajectories in non-straight electron trajectories. Thus, around this $\boldsymbol{R}(\phi_P, \phi)$, the ether points rotate and these axes of rotation are parallel to the motion of this $\boldsymbol{R}(\phi_P, \phi)$. These rotations of ether points form the field of rotation around the trajectory of $\boldsymbol{R}(\phi_P, \phi)$. The perturbations caused by this motion is propagated in the free ether with velocity c and is a propagation of light, since it is created by $\boldsymbol{R}(\phi_P, \phi)$ **of zero mass and also of null electrical charge, that is, created by** $\boldsymbol{R}(\phi_P, \phi)$**, which is a photon**.

IV.8 Recapitulation of Sec. IV

I have shown that Maxwell's electrodynamics ensue from the ether elasticity equation, namely, equation (3.5), from which the following physical parameters and properties follow.

An electromagnetic field is a change in the ether defined as follows. The magnetic field $\boldsymbol{H}$ denotes the speed of the rotations of the ether points (Cf. Eq. (4.2)); the magnetic induction $\boldsymbol{B}$ is the volumetric density of the linear momentum of the ether points; and the electric field $\boldsymbol{E}$ is defined by Eq. (4.1). The electric charges and currents ensue (Cf. eqs. (4.4) and (4.5)) from a volumetric density of couples $\boldsymbol{C}$ applied to the ether. The forces to which an electric charge in an electromagnetic field result from the interactions of changes in the ether. The magnetic force results from the interaction of two magnetic fields: the one created by the electric charge and the incident one. The magnetic field created by the charge in question is, in its near vicinity, the same as a vortex flow in a fluid. Thus, the interaction of the two magnetic fields is analogous with what happens in a fluid when a vortex flow and an incident linear flow interact. The electric force derives from the electromagnetic energy of a closed system of electrically charged particles.

$\boldsymbol{R}(\phi_P, \phi)$, defined in (4.45), and called a single-photon wave, represents a photon in which the amplitude is non-zero only in the volume defined by $SINC\left[\frac{\Delta\omega}{2}\left(-t+\frac{x}{c}\right)\right]$, which moves at velocity c.

CHAPTER V

THE GENERALIZATION OF ELASTIC ELECTROMAGNETISM AND RESULTS ENSUING FROM THE LAGRANGE-EINSTEIN FUNCTION

V.1 Introduction to the Generalization of Elastic Electromagnetism

In discussing the elastic interpretation of electromagnetism, I introduced the notion of the density of couples of forces $\boldsymbol{C}$ applied to the ether. These couples of forces $\boldsymbol{C}$ are such that we can deduce from them the electrical charges e and currents $\boldsymbol{j}_e$ that create the electromagnetic field. However, this field is constituted by photons and therefore it is $\boldsymbol{C}$ that generates the photons and the electromagnetic field. This field, and therefore the photons, can exist even in a domain where $\boldsymbol{C} = 0$, i.e., one that is void of electrical charges and currents, as shown, for example, in Sec. IV.5. Let us consider again the electromagnetic field in a spatial domain where $\boldsymbol{C} = 0$, and where the ether is free, i.e., in the absence of gravitation, but in the case where this electromagnetic field, which is a monochromatic wave, is present. As demonstrated above, this electromagnetic wave is formed by photons and photons are reciprocally constituted by electromagnetic waves. However, these waves, i.e., these photons, were created by couples of forces $\boldsymbol{C}$ applied to the ether somewhere outside of the free domain of the ether in which one considers the propagation of such electromagnetic monochromatic waves. It follows that

> a density of couples of forces $\boldsymbol{C}$ is also, a generator of electromagnetic waves, i.e., of photons that are propagated in the ether even where $\boldsymbol{C}$ is absent.

Since $\boldsymbol{C}$ is a generator of photons, i.e., of $Par(0{,}0)s$, it can also be denoted more generally $G(0{,}0)$. A spatial domain in the ether containing $Par(0{,}0)s$ and the generator (of photons) $G(0{,}0)$ will be denoted $D[G(0{,}0), Par(0{,}0)s]$, while a spatial domain in the ether containing $Par(0{,}0)s$, but not the generators $G(0{,}0)$ will be denoted $[0, Par(0{,}0)s]$.

We can generalize the concept of the photon generator to the concept of the massive particle generator, which can be electrically charged, i.e. the concept of the generator of $Par(m,e)s$. In the same way that $G(0,0)s$ arises, i.e., photon generators of $Par(0,0)s$, we can conceive that $Par(m,e)s$ is also created by generators that I denote $G(m,e)s$. By analogy with $D[G(0,0), Par(0,0)s]$ and $D[0, Par(0,0)s]$, I denote by $D[G(m,e), Par(m,e)]$ a spatial domain in the ether containing $Par(m,e)s$ and the generators $G(m,e)s$, and by $D[0, Par(m,e)s$ a spatial domain in the ether containing $Par(m,e)s$, but not the generators $G(m,e)s$. We should note that $\mathfrak{R}(g,A)$ denotes a spatial domain in which a gravitation field and an electromagnetic field exist. I go on to consider what happens in $D[0, Par(m,e)s$ and in $\mathfrak{R}(g,A)$, that is to say, in $D[0, Par(m,e)s \cap \mathfrak{R}(g,A)$.

Electromagnetism ensues from the wave $\boldsymbol{\xi}_{EM}$, denoted $\boldsymbol{\xi}$ for simplicity and defined in Eq. (3.5), from which we can deduce Maxwell's equations (Cf. Sec. IV.1). This particular case considers that the phase velocity in the free ether is c, defined by $c^2 = \eta_0/\rho_0$, which is the free light velocity in the ether. For the wave propagated there at velocity c, that is, for the given constant ρ_0, the coefficient η defines the free light velocity when it takes the value η_0. Such a wave $\boldsymbol{\xi}$, that can be denoted $\boldsymbol{\xi}_{EM}$, is created by couples of forces denoted $\boldsymbol{C}$ (one should not confound c and $\boldsymbol{C}$), applied to the ether. This wave is an electromagnetic one that propagates in the ether at the phase velocity, which, in this case, is also the group velocity c. A superposition of such waves of close frequencies forms a "rotating ether trajectory segment", called a photon, that also moves in the free ether at velocity c. Therefore, a distribution of these couples $\boldsymbol{C}$ applied to the ether creates waves the specific superposition of which build photons. These waves are solutions of Eq. (3.5) and are defined by the fact that, in the ether, only a distribution of couples of forces $\boldsymbol{C}$ is applied, such that, in the expression for the light velocity, it is the coefficient η_0. This is generally denoted η, which, for the given ρ_0, defines the phase velocity, being the free light velocity c in the free ether. Therefore, the distribution $\boldsymbol{C}/(2\eta_0)$ creates in the ether a wave and/or a superposition of such waves creating a "rotating ether trajectory segment" that also moves at velocity c. As stated above, these waves are the solutions of Eq. (3.5).

In order to generalize equation (3.5), which gives the fundamental equation of electromagnetism, we must consider the case in which the **phase** velocity in the ether is **not** c, but rather $\boldsymbol{V}_P$, defined by $V_P{}^2 = \eta/\rho_0$. The wave $\boldsymbol{\xi}$ propagates in the ether at the phase velocity $\boldsymbol{V}_P$, that is, for the given ρ_0, the coefficient η defines the particle velocity that is now different to η_0, the wave being created, not by $\boldsymbol{C}/(2\eta_0)$, but now by $\boldsymbol{C}/(2\eta)$, where $\boldsymbol{C}$ is a couple of forces applied to the ether and where η_0 becomes η. This wave propagates in the ether at the **phase** velocity $\boldsymbol{V}_P$, which, in this case, is different from the **group** velocity $\boldsymbol{V}$, as shown in the following, and, a superposition of waves of phase velocities $\boldsymbol{V}_P$ of close frequencies forms a "rotating ether trajectory segment" called a particle, which moves **not** at the phase velocity $\boldsymbol{V}_P$, but at the group velocity $\boldsymbol{V}$. Therefore, a distribution of $\boldsymbol{C}/(2\eta)$ creates waves that can form a superposition, building massive particles. In the domain of $\boldsymbol{C} = 0$, these waves are solutions of Eq. (3.7) that can be written as in eqs. (3.8) or (3.9), defined by the fact that a distribution of couples of forces divided by the 2η that is by $\boldsymbol{C}/(2\eta)$ is applied in the ether, such that it is the coefficient η that, for the given ρ_0, and for a given $\boldsymbol{C}$ that defines the phase velocity $\boldsymbol{V}_P$ associated with this particle. Thus, the distribution $\boldsymbol{C}/(2\eta)$ creates a wave and or a superposition of such waves in the ether, creating a "rotating ether trajectory segment" moving at velocity $\boldsymbol{V}$ (which is different to $\boldsymbol{V}_P$), In the domain where $\boldsymbol{C}$ is absent, these waves are generally the solutions of Eq. (3.7).

It is a fact that the waves $\boldsymbol{\xi}_{0,0}$ (denoted also simply $\boldsymbol{\xi}$ in the absence of confusion) that propagate in the ether and that are the solution of Eq. (4.29), are associated with $Par(0{,}0)s$, and waves $\boldsymbol{\xi}_{m,e}$ (denoted also simply $\boldsymbol{\xi}$ in the absence of confusion) that also propagate in the ether are associated with $Par(m,e)s$. These last $\boldsymbol{\xi}$ are the solutions of an equation that generalizes Eq. (4.29), taking into consideration the incident gravitational and electromagnetic fields, since an $\boldsymbol{\xi}_{m,e}$ is constituted by $Par(0{,}0)s$ in $D[0, Par(m,e)s \cap \Re(g,A)$. Note that that $Par(m,e)$ becomes $Par(0{,}0)$ when $m = 0$, and that $Par(0{,}0)$ and the wave $\boldsymbol{\xi}_{0,0}$ are influenced only by gravitational fields.

In $\Re(g,A)$, the equation that generalizes Eq. (4.29), differs from it due to the fact that light velocity c is generalized by phase velocity $\boldsymbol{V}_P$ of the wave $\boldsymbol{\xi}_{m,e}$. This phase velocity $\boldsymbol{V}_P$ ensues from the fact that, as demonstrated below, the phase ϕ of this $\boldsymbol{\xi}_{m,e}$ ensues from the Lagrange-Einstein function.

Two mechanical concepts, namely, the undulating one and the probabilistic one, appear contradictory and even de Broglie changed his opinion about them. On fields and particles, Einstein wrote the statement cited in Sec. I.1. By generalizing Sommerfeld's results, I demonstrated in Chap. IV that electrodynamics is identical to a case of the elasticity theory called the *elastic interpretation of electrodynamics*. This result encouraged us to search for the common physical nature of fields and particles that suppresses the contradiction and inconsistency cited here by Einstein. It is natural to try and extend this elastic interpretation of electrodynamics to the gravitational field and to particles subject to it, including photons.

This extension is made possible by considering the Lagrange-Einstein function L_G of a particle of rest mass m and electric charge e, subjected to both a gravitational and an electromagnetic field. We find that L_G not only yields particle motion, but is such that ϕ, defined by $\hbar d\phi/dt = L_G$, is the phase of a wave $\boldsymbol{\xi}$. General relativity implies, therefore, that the waves are associated with the particle. Since these waves are propagated in an elastic medium, it follows that $\boldsymbol{\xi}$ is also propagated in an elastic medium. However, as shown in Sec. IV for $m = 0$, $\boldsymbol{\xi}$ is an electromagnetic wave. Therefore, the elastic medium in which $\boldsymbol{\xi}$ is propagated is the ether. Maxwell considered that the existence of the ether (see quotation in Sec. I.1. Lorentz, in Ref. 18) explained the Zeeman effect. Some people consider that Einstein ruled out the existence of the ether. This is not quite right since he simply derived Lorentz's transformation from a mathematical procedure that was very elegant and explained why Lorentz's construction was independent of the material of the experiment. Lorentz wished to explain how Michelson's result could occur in the presence of the ether that Michelson-Morley was supposed to measure. Therefore, the existence of the ether was never really disproved, but rather left in question, and we can thus make the assumptions discussed here. Einstein's words (Quot. 3) confirm this.

Let us consider now the "classical particle", by definition being a particle the movement of which is univocally determinable given adequate initial conditions and the fields to which it is subject. Considering Einstein's words, it appears that the particle must be something other than a simple material point that is discontinuous in space. In this context, I demonstrate that, as with the wave-photon property, a sum exists, denoted $\boldsymbol{R}(\phi_P, \phi)_{\boldsymbol{m,e}}$, of waves $\boldsymbol{\xi}_{m,e}$ (denoted also simply $\boldsymbol{R}(\phi_P, \phi)$ of waves $\boldsymbol{\xi}$ in the absence of confusion), forming a "rotating ether trajectory segment". This is, in fact, a particle, being *a "shock wave" that moves like a particle.* It is not dispersed and occupies a limited spatial region containing the particle energy. The classical particle is, in fact, such a globule.

> *A classical particle is a rotating ether trajectory segment* $\boldsymbol{R}(\phi_P, \phi)$ *constituted by an integral of waves* $\boldsymbol{\xi}$. *When* $m = 0$, *the particles are photons, and the waves are electromagnetic.*

This property of wave-particle reciprocity suppresses the inconsistency described by Einstein. It also suppresses the apparent contradictory aspects of the two mechanical concepts—the probabilistic one and the undulating one—as we will see more precisely in Sec. IX. Indeed, I show that the univocal motion of the classical particle (massive or not) is due to the fact that it is a full superposition of waves that forms this $\boldsymbol{R}(\phi_P, \phi)$. Such a full superposition can exist only in an unbound state. In a bound state, the waves that constitute $\boldsymbol{R}(\phi_P, \phi)$ interfere with each other and only a few of them, the resonant ones, remain undestroyed by these interferences and may even be amplified. These resonant waves, called "quantum states", can be directly determined even in curved space times (Cf. Sec. IX). When $m = 0$, this $\boldsymbol{R}(\phi_P, \phi)$ is a sum of electromagnetic waves that behave like a hard projectile, causing, for example the photo-electric effect.

V.2 The Lagrange-Einstein Function and the Lagrange Form of Einstein's Particle Motion Equation

I show now that the Lagrange-Einstein function L_G of a $Par(m, e)$ subject to a gravitational field $g_{\mu\nu}$ and an electromagnetic potential tensor A_μ, being

$$L_G = -mc\dot{s} + eA_\mu \dot{x}^\mu / c \tag{5.1}$$

the Einstein particle motion equation

$$\frac{d^2 x^\sigma}{ds^2} = \Gamma^\sigma_{\mu\nu} \frac{dx^\mu}{ds} \frac{dx^\nu}{ds} \tag{5.2}$$

where $\Gamma^\sigma_{\mu\nu}$ are the Christoffel symbols can be written in the following Lagrange form of the motion equation

$$\frac{d}{dt}\left(\frac{\partial L_G}{\partial \dot{x}^\mu}\right) - \partial_\mu L_G = 0\ . \tag{5.3}$$

Indeed, we can already see that (5.1) generalizes Eq. (16.4) of Ref. 16. In order to demonstrate eqs. (5.1) and (5.3), let us first consider the case where e = 0. On account of (1.67), we have the two following relations

$$\partial \dot{s} / \partial \dot{x}^\alpha = g_{\mu\alpha} dx^\mu$$

$$\partial_\mu \dot{s} = (1/2) \partial_\mu g_{\alpha\beta} dx^\alpha \dot{x}^\beta / ds\ .$$

The insertion of these two relations into (5.3) gives

$$\frac{d}{ds}\left(g_{\mu\alpha} \frac{dx^\sigma}{ds}\right) - \frac{1}{2}\left(\partial_\mu g_{\alpha\beta}\right) \frac{dx^\alpha}{ds} \frac{dx^\beta}{ds} = 0 \tag{5.4}$$

which, after the differentiation of the left-hand member and taking into account the identity,

$$\partial_\nu g_{\alpha\mu} dx^\mu dx^\nu = \partial_\mu g_{\alpha\nu} dx^\nu dx^\mu$$

becomes

$$g_{\alpha\sigma}\frac{d^2x^\sigma}{ds^2} + \frac{1}{2}\left(\partial_\nu g_{\alpha\mu} + \partial_\mu g_{\alpha\nu} - \partial_\alpha g_{\mu\nu}\right)\frac{dx^\mu}{ds}\frac{dx^\nu}{ds} = 0\,. \quad (5.5)$$

After the contracted multiplication of this equation by $g^{\tau\alpha}$, we can see that Eq. (5.3) becomes Eq. (5.2), which is the particle motion equation (46) of Ref. 19. Thus, I have proven that the Lagrange-Einstein function L_G of a neutral particle in a gravitational field is $-mc\dot{s}$, from which one deduces . immediately, considering (5.3), where generally, even for $e \neq 0$, the Lagrange-Einstein function of this particle is given by Eq. (5.1), and that its motion equation can thus take the form (5.3).

V.3 The Particle Momentum-energy Tensor Ensuing from the Lagrange-Einstein Function

The expression for the momentum tensor P_μ is defined by

$$P_\mu \equiv \partial L_G / \partial \dot{x}^\mu \quad (5.6)$$

which, considering (5.1), is

$$P_\mu = -mc dx_\mu / ds + eA_\mu / c\,. \quad (5.7)$$

From (5.1), (5.7), and $dx^\mu dx_\mu = (ds)^2$, it follows that

$$L_G dt = P_\mu x^\mu = P_4 x^4 + P_i x^i\,. \quad (5.8)$$

I shall show that the total particle energy denoted E_T is constant when the fields to which the particle is subjected are time-independent and can always be written in the form

$$E_T = mc^2 + h\nu \quad (5.9)$$

where ν has the dimension of a frequency and h is the Planck constant. Thus, $h\nu$ represents the part of the particle energy that is not the massive-particle rest energy, when $m = 0$, then $h\nu$ represents the photon energy.

When the incident fields are time independent, then, considering the expression of L_G given in Eq. (5.1), we can see that $\partial_t L_G = 0$. Setting $\mu = 4$ in Eq. (5.3) and in Eq. (5.6), we get $dP_4/dt = 0$, that is, P_4 is constant. P_4 is related to the particle total energy E_T by the relation

$$P_4 = -E_T/c \,. \tag{5.10}$$

Therefore, when the fields to which the particle is subjected are time independent, E_T is constant, since P_4 is also constant. On account of the relation $dx_4 = g_{4\mu}dx^\mu$ of eqs. (5.7), (5.9), and (5.10), we find generally

$$E_T \equiv mc^2 + h\nu \equiv -cP_4 = mc^2 g_{4\mu}dx^\mu/ds - eA_4 \tag{5.11}$$

which, for the photon, i.e., for $m = 0$, becomes

$$E_T \equiv h\nu \equiv -cP_4 = c^2 g_{4\mu}dx^\mu \lim_{m\to 0}(m\,/ds)\,. \tag{5.12}$$

This limit can exist since, as shown by Einstein, ds tends toward zero when m tends toward zero.

In stationary fields, then, since one can choose the coordinates to be such that $g_{4j} = 0$, these two equations become, respectively

$$E_T \equiv mc^2 + h\nu = mc^3 g_{44}/\dot{s} - eA_4 \tag{5.13}$$

and

$$E_T \equiv h\nu = c^3 g_{44} \lim_{m\to 0}(m\,/\dot{s})\,. \tag{5.14}$$

V.4 The Phase and Phase Velocity of the Particle Wave Ensuing from the Lagrange-Einstein Function

Let ϕ_P denote the quantity defined by

$$d\phi_P = L_G dt/\hbar \tag{5.15}$$

that is, on account of (5.8) and (5.10),

$$\phi_P = \left(-\int E_T dt + \int P_j dx^j\right)/\hbar \qquad (5.16)$$

where the integral is taken along a geodesic defined by Eq. (5.3). Equation (5.16) can also be written

$$\phi_P = \left(-\int E_T\, dt + \int P_j\, u^j d\ell\right)/\hbar \qquad (5.17)$$

i.e.,

$$u^j = dx^j/d\ell$$

and $E_T/\hbar$ has the dimensions of a frequency. Therefore, ϕ is the phase of a wave ξ of the form

$$\xi = \xi_0 exp(i\phi_P) \qquad (5.18)$$

called the "classical particle wave" and defined in Sec. VI.2. We can say that the ray of this wave is a particle trajectory determined from the motion equation (5.3) and where ξ_0 is taken as a constant. The phase velocity V_P of the phase ϕ_P of ξ is defined by the fact that $\phi_P = 0$ on a ray when

$$d\ell = V_P dt$$

i.e., when

$$\int(-E_T + P_j u^j V_p)dt = 0 \qquad (5.19)$$

that is to say when

$$V_p = E_T/\left(P_j u^j\right) \qquad (5.20)$$

the explicit expression for V_p being given in Sec. V.5 below. Expression (5.18) can be written

$$\xi = \xi_0 exp\left[\frac{i}{\hbar}\left(-\int E_T dt + \int \frac{E_T}{V_P} d\ell\right)\right] \qquad (5.21)$$

which, in the time independent case, i.e., in a stationary field, becomes

$$\xi = \xi_0 exp\left[\frac{i}{\hbar}E_T\left(-t + \int \frac{d\ell}{V_P}\right)\right]. \qquad (5.22)$$

In Sec. VI.2, I show that the wave ξ defined in (5.21) is the solution of a generalized ether elasticity equation (4.29).

V.5 Explicit Expressions for Particle Velocity and Phase Velocity

From (5.7) and (5.10), we can deduce

$$mc^2/ds = (E_T + eA_4)/dx \tag{5.23}$$

and from (5.7) and (5.23), we obtain

$$P_j = \frac{(E_T+eA_4)}{c}\frac{dx_j}{dx_4} + \frac{eA_j}{c}\,. \tag{5.24}$$

Using the notation

$$b \equiv (E_T + eA_4)/(mc^2) \tag{5.25}$$

(1.7) and (5.23) yield

$$\left(g_{4\mu}g_{4\nu} - b^2 g_{\mu\nu}\right)dx^\mu dx^\nu = 0\,. \tag{5.26}$$

If G and B denote the quantities defined by

$$G \equiv \sqrt{u^i u^j (g_{j4}g_{i4} - g_{ij}g_{44})} \tag{5.27}$$

$$B = \sqrt{1 - b^{-2}g_{44}} \tag{5.28}$$

then, by solving (5.26) for u^4, and considering that

$$u^4 \equiv cdt/d\ell \equiv c/V$$

we obtain

$$u^4 = \left(G/B - g_{j4}u^j\right)/g_{44}\,. \tag{5.29}$$

We can deduce the explicit expression for the velocity V of a particle subjected to a gravitational and an electromagnetic field, namely

$$V = cg_{44}B/\left(G - Bg_{j4}u^j\right) . \tag{5.30}$$

Furthermore, since

$$x_4 \equiv g_{4\mu}dx^\mu \equiv g_{44}dx^4 + g_{4j}dx^j$$

then, from (5.29) we also have

$$u_4 = G/B . \tag{5.31}$$

Inserting (5.31) into (5.24), we obtain

$$P_j = -\frac{(E_T+eA_4)}{c}\frac{B}{G}u_j + \frac{eA_j}{c} \tag{5.32}$$

but

$$u_j = g_{jk}u^k + g_{j4}u^4 . \tag{5.33}$$

Therefore, on account of (5.29), Eq. (5.33) can be written

$$u_j = [(g_{44}\, g_{ik-} - g_{j4}\, g_{k4})\, u^k + g_{j4}\, G/B]/g_{44} . \tag{5.34}$$

Inserting this expression for u_j into (5.32), we obtain the following expressions for the spatial components p_j of the momentum tensor and for $p_j u^j$ on a ray

$$p_j = \frac{eA_j}{c} - \frac{1}{cg_{44}}(E_T + eA_4)\left[\frac{B}{G}\left(g_{44}\, g_{ik} - g_{j4}\, g_{k4}\right)u^k + g_{j4}\right] \tag{5.35}$$

that is to say,

$$p_j u^j = \frac{eA_j u^j}{c} - \frac{1}{cg_{44}}(E_T + eA_4)[GB - g_{j4}u^j\,) . \tag{5.36}$$

From (5.20) and (5.36), we can deduce that the phase velocity expression V_P of $\boldsymbol{\xi}$ on a ray, is

$$V_P = \frac{cE_T g_{44}}{eA_j u^j g_{44} + (E_T+eA_4)[GB - g_{j4}u^j\,)} . \tag{5.37}$$

The explicit expression for the wave ξ on a ray is obtained, therefore, by inserting Eq. (5.37) into Eq. (5.21). In the case of $m = 0$, the expressions containing m bear $m = 0$ as a low index, e.g., from (5.25) and (5.28), we have $B_{m=0} = 1$. Therefore, since $e = 0$ for a photon, then (5.37) yields

$$V_{P,m=0} = \frac{cg_{44}}{G - g_{j4}u^j} \,. \tag{5.38}$$

Considering (5.30), it follows that

$$V_{m=0} = V_{P,m=0} \,. \tag{5.39}$$

Remembering that, here also, I denote ϕ as the "phase group", the quantity defined by Eq. (4.43).

V.6 Relations Connecting the Particle Wave Phase, the Phase Velocity, the Group Phase, and the Group Phase Velocity

Note that the group phase is defined in (4.43), i.e., by

$$\phi = -t + \int \frac{\partial}{\partial E_T}\left(\frac{E_T}{V_P}\right) d\ell \tag{5.40}$$

where ϕ is the phase of ξ defined in (5.21) or in (5.22), that is

$$\phi_P = \frac{1}{\hbar}\left(-\int E_T dt + \int \frac{E_T}{V_P} d\ell\right) \tag{5.41}$$

which, for the constant E_T, becomes simply

$$\phi_P = \omega\left(-t + \int \frac{1}{V_P} d\ell\right). \tag{5.42}$$

From (5.37), after simple calculations, we have

$$\frac{\partial}{\partial E_T}\left(\frac{E_T}{V_P}\right) = \frac{1}{cg_{44}B}\left[G - Bg_{j4}u^j\right) \tag{5.43}$$

comparing the right-hand member of (5.43) and (5.30), we derive the very important relation (1.4) and therefore (5.40) becomes

$$\phi = -t + \int d\ell/V \tag{5.44}$$

which generalizes Eq. (4.44). We can see that that the group phase velocity is simply the particle velocity V.

As shown in Sec. V.2, the Einstein four-motion equation of $Par(m,e)$ in $\Re(g,A)$ can be formulated in the Lagrange form defined in Eq. (5.3). In addition, the expression for the momentum tensor P_μ defined by Eq. (5.6) is given in Eq. (5.7), from which follows the relation (5.8). Furthermore, $-cP_4$ is the total particle energy denoted E_T, for which the expression is given in (5.11), and for the photon, E_T becomes simply $h\nu$. However, from eqs. (5.15), (5.18), and (5.21), we can obtain Eq. (5.41), where V_P denotes the phase velocity of a wave ξ that generalizes the one defined by eqs. (4.30)-(4.32). This wave ξ of phase velocity V_P, as defined above, is of phase ϕ where the spatial curvilinear integral is taken along a trajectory defined in Sec. V.7-A. The expression for V_P of ξ associated with $Par(m,e)$ in $\Re(g,A)$ is defined in Eq. (5.37) and by eqs. (5.25), (5.27), and (5.28). I note also that the expression for the particle velocity V is given in (5.30) and that V is related to V_P by Eq. (1.4). Therefore, if we define ϕ by (4.43), then, on account of (5.41) and (1.4), we obtain (5.44). In $\Re(g,A)_0$, V and V_P are obtained from (5.30) and (5.37) by setting $A_j = 0$ and $g_{4j} = 0$, and then obtaining the expressions given in Sec. V.8.

V.7 Further Results Ensuing from the Lagrange-Einstein Function

A. Trajectory equation of $Par(m,e)$ in $\Re(g,A)$ generalizing the light ray equation

From eqs. (5.15), (5.37) and (5.41), it follows that the Lagrange-Einstein L_G of $Par(m,e)$ in $\Re(g,A)$ can also be written

$$L_G = E_T(-1 + \dot{\ell}/V_P) \tag{5.45}$$

where $\dot{\ell}$ denotes the path length covered per unit of time, on the wave ray, i.e., on the particle trajectory. Let us now consider the case in which the particle is in $\Re(g,A)$, since E_T is constant, then (5.45) and (5.3) yields

$$\frac{d}{dt}\frac{\eth}{\partial \dot{x}^j}\left(\frac{\dot{\ell}}{V_{P0}}\right) = \eth_j\left(\frac{\dot{\ell}}{V_{P0}}\right). \tag{5.46}$$

Our purpose here is to obtain the trajectory of the particle, i.e., to obtain $u^j = u^j(\ell)$ and $x^i = x^i(\ell)$. However, on this trajectory, V_{P0} is a function of $x^i(\ell)$, from which we can deduce $u^j(\ell)$, i.e., $V_{P0} = V_{P0}[x^i(\ell)]$, which can be written simply $V_{P0}(\ell)$. Furthermore, $\dot{\ell}$ is a function of only the $\dot{x}^i$, therefore, Eq. (5.46), can be written

$$\frac{d}{d\ell}\left[\frac{1}{V_{P0}(\ell)}\frac{\partial\dot{\ell}}{\partial\dot{x}^j}\right] = \frac{\partial}{\partial x^i(\ell)}\left[\frac{1}{V_{P0}(\ell)}\right]. \tag{5.47}$$

Considering that, in Cartesian coordinates

$$\dot{\ell} = \sqrt{\Sigma_{i=1}^{3}(\dot{x}^i)^2}$$

and so it follows that

$$\frac{\partial\dot{\ell}}{\partial\dot{x}^j} = \frac{dx^j}{d\ell} = u^j(\ell) \tag{5.48}$$

therefore, inserting (5.48) in (5.47), we have

$$\frac{d}{d\ell}\left[\frac{1}{V_{P0}(\ell)}u^j(\ell)\right] = \frac{\partial}{\partial x^i(\ell)}\left[\frac{1}{V_{P0}(\ell)}\right]. \tag{5.49}$$

Eq. (5.49) is the same as the optical differential equation of light rays (Cf. Sec. 3.2 of Ref. 20).

B. The elastic interpretation of Einstein's infinitesimal element ds

Considering eqs. (5.45) and (5.1), it follows that

$$mc\dot{s} = E_T\left(1 - \dot{\ell}/V_P\right) + eA_\mu\dot{x}^\mu/c\,. \tag{5.50}$$

Denoting by $F_{\mu\nu}$ the tensor defined by

$$F_{\mu\nu} = \partial_\mu A_\nu - \partial_\nu A_\mu \tag{5.51}$$

and considering (4.1) and (4.3), we can verify that

$$\dot{x}^i F_{ij} = -(\boldsymbol{V}\wedge \boldsymbol{curlA})_j = (\boldsymbol{V}\wedge\boldsymbol{B})_j = (\boldsymbol{V}\wedge\rho_0\partial\boldsymbol{\xi})_j \tag{5.52}$$

$$F_{4j} = E_j = \eta(\boldsymbol{curl\xi})_j \,. \tag{5.53}$$

It follows that $F_{\mu\nu}$, i.e., $\boldsymbol{A}_\mu$, are functions of $\boldsymbol{\xi}$, and therefore, as shown in (5.50), Einstein's element ds is also a function of $\boldsymbol{\xi}$; this is the elastic interpretation of ds.

V.8 Example Cases Where the Particles are Subjected to Time-independent Fields

V.8.1 Generalities of the case where the particle is subjected to a time-independent field

In time-independent fields, the total particle energy $E_T = mc^2 + h\nu$ is constant, $\boldsymbol{A}_j = 0$, and we can choose the coordinates to be such that $g_{4j} = 0$. With these conditions, (5.27) becomes

$$G \equiv \sqrt{-u^i u^j g_{ij} g_{44}} \tag{5.54}$$

and (5.25), (5.28), (5.30), and (5.37), yield

$$V = \frac{c g_{44} B}{G} \tag{5.55}$$

$$V_P = \frac{c E_T g_{44}}{(E_T + eA_4) G B} \,. \tag{5.56}$$

For a neutral massive particle, we can set e = 0 in (5.55) and (5.56). Furthermore, for a photon, we set $m = 0$ in these equations and obtain

$$V_{m=0} = V_{P,m=0} = c g_{44}/G \,. \tag{5.57}$$

For a particle in an EM field only, we set

$$g_{44} = 1,\ G = 1, \text{ and } B = \sqrt{1 - \left[\frac{mc^2}{(E_T + eA_4)}\right]^2}$$

in (5.55) and (5.56), while (5.57) becomes

$$\mathrm{V}_{\mathrm{m=0}} = \mathrm{V}_{\mathrm{p,m=0}} = \mathrm{c} \,.$$

If the particle is free, then the expression for ξ, considering that the ray is the x axis, for example, is

$$\xi = \xi_0 exp\left[\frac{i}{\hbar}\left(-E_T t + \frac{x}{c}\sqrt{{E_T}^2 - (mc^2)^2}\right)\right] \tag{5.58}$$

the expression for V, denoted more specifically by V_∞, is then

$$V_\infty = c\sqrt{1 - (mc^2/E_T)^2}\ . \tag{5.59}$$

For $m \gg h\nu/c^2$, i.e., in the non-relativistic case, (5.58) gives the de Broglie plane wave, which, for $m = 0$, is the electromagnetic wave

$$\xi = \xi_0 exp[i2\pi\nu(-t + x/c)]\ . \tag{5.60}$$

V.8.2 The case in which the particle is subjected to a Schwarzschild gravitational field

If the neutral particle is subject to a Schwarzschild gravitational field, then

$$\begin{aligned} x^1 &= r \\ x^2 &= \theta \\ x^3 &= \varphi \\ x^4 &= ct \end{aligned} \tag{5.61}$$

where θ denotes the angle made by the z axis and the particle radius vector $\boldsymbol{r}$ and φ denotes the angle made by the x axis and the projection of $\boldsymbol{r}$ on the plane $z = 0$. Furthermore, by m_0 we can denote the immobile mass that creates the Schwarzschild metric, the constant of gravitation by k, and by α the quantity defined by

$$\alpha \equiv 2m_0 k/c^2 \tag{5.62}$$

$\widehat{Vr}$ concerns the angle made by the velocity $\boldsymbol{V}$ and $\boldsymbol{r}$, and γ^2 and $\hat{\gamma}^2$ denote the quantities defined by

$$\gamma^2 = 1 - \alpha/r$$

$$\hat{\gamma}^2 = 1 + \alpha(cos^2\widehat{Vr})/(r\gamma^2)\ . \tag{5.63}$$

Noting that the non-zero $g_{\mu\nu}$ of this field are

$$g_{11} = -1/\gamma^2$$
$$g_{22} = -r^2$$
$$g_{33} = -r^2 sin^2\theta$$
$$g_{44} = \gamma^2 \tag{5.64}$$

we can verify that ds can then be written in the form

$$ds = \sqrt{(cdt\gamma)^2 - (d\ell\hat{\gamma})^2} \tag{5.65}$$

and that

$$\hat{\gamma}^2 = -u^i u^j g_{ij} \tag{5.66}$$

$$B = \sqrt{1 - (\gamma\, mc^2/E_T)^2}\ . \tag{5.67}$$

Therefore, on account of (5.55), (5.56), (5.63), and (5.67), the expressions for the particle velocity V and for the phase velocity V_P of the particle wave are

$$V = cB\gamma/\hat{\gamma} \tag{5.68}$$

$$V_P = c\gamma/(\hat{\gamma}B)\ . \tag{5.69}$$

For $\mathrm{m} = 0$, (5.68) and (5.69) become

$$V_{m=0} = V_{P,m=0} = c\gamma/\hat{\gamma}\ . \tag{5.70}$$

Now, we have

$$(ds)^2 = g_{11}(dr)^2 + g_{22}(d\theta)^2 + g_{33}(d\varphi)^2 + g_{44}(dct)^2 \tag{5.71}$$

being, in this field

$$(ds)^2 = -\frac{1}{\gamma^2}(dr)^2 - r^2(d\theta)^2 - r^2 sin^2\theta(d\varphi)^2 + \gamma^2(dct)^2$$

that is,

$$\left(\frac{ds}{dt}\right)^2 = -\frac{1}{\gamma^2}\dot{r}^2 - r^2\dot{\theta}^2 - r^2 sin^2\theta\dot{\varphi}^2 + \gamma^2 c^2\ . \tag{5.72}$$

Furthermore, the general expression for the velocity is such that

$$V^2 = \dot{r}^2 + \left(r\dot{\theta}\right)^2 + r^2\dot{\varphi}^2 sin^2\theta \qquad (5.73)$$

so that

$$V^2 - \dot{r}^2 = \left(r\dot{\theta}\right)^2 + r^2\dot{\varphi}^2 sin^2\theta$$

and

$$\left(\frac{ds}{dt}\right)^2 = \gamma^2c^2 - \frac{\dot{r}^2}{\gamma^2} - (V^2 - \dot{r}^2) = \gamma^2c^2 - V^2 + \dot{r}^2\left(1 - \frac{1}{\gamma^2}\right) \qquad (3.74)$$

Considering

$$\dot{r} \equiv V cos\left(\widehat{\boldsymbol{Vr}}\right). \qquad (5.75)$$

Recalling that $\widehat{\boldsymbol{Vr}}$ is the angle made by the radius vector **r** and the velocity **V**, then

$$\left(\frac{ds}{dt}\right)^2 = \gamma^2c^2 - V^2 + V^2\left(cos^2\widehat{\boldsymbol{Vr}}\right)\left(1 - \frac{1}{\gamma^2}\right). \qquad (5.76)$$

After a small calculation, and considering Eq. (5.63), we have

$$\left(\frac{ds}{dt}\right)^2 = \gamma^2c^2 - \hat{\gamma}^2V^2 . \qquad (5.77)$$

The general expression (5.77) for $\left(\frac{ds}{dt}\right)^2$ of a neutral massive particle in a Schwarzschild field, can be written, considering Eq. (5.68), as follows

$$\left(\frac{ds}{dt}\right)^2 = \left(c\gamma^2\frac{mc^2}{E_T}\right)^2 . \qquad (5.78)$$

Taking the particle trajectory in the plane defined by $\theta = \pi/2$, then

$$V^2 = \dot{r}^2 + r^2\dot{\varphi}^2 \qquad (5.79)$$

and the expressions for V and V_P are given in eqs. (5.68) and (5.69). In addition $\dot{r}^2 = 0$, i.e., for $V^2 = r^2\dot{\varphi}^2$, that is, for a circular trajectory, i.e., for $\boldsymbol{Vr} = 0$, then $\hat{\gamma} = 1$, and eqs. (5.68) and (5.69) become

$$V = c\gamma\sqrt{1-\left(\gamma\frac{mc^2}{E_T}\right)^2} \tag{5.81}$$

$$V_P = c\gamma\frac{1}{\sqrt{1-\left(\gamma\frac{mc^2}{E_T}\right)^2}}. \tag{5.81}$$

V.8.3 A massive and electrically charged particle in a Schwarzschild field and in the field created by the electrically charged particle

In this case, B, defined in (5.67), becomes B_e defined by

$$B_e = \sqrt{1-\left(\frac{\gamma}{b_e}\right)^2} \tag{5.82}$$

where

$$b_e \equiv \frac{E_T+eA_4}{mc^2} \tag{5.83}$$

that is,

$$B_e = \sqrt{1-\left(\frac{\gamma mc^2}{E_T+eA_4}\right)^2} \tag{5.84}$$

and V and V_P become V_e and $V_{e,P}$, for which the expressions are

$$V_e = cB_e\gamma/\hat{\gamma} \equiv c\frac{\gamma}{\hat{\gamma}}\sqrt{1-\left(\frac{\gamma mc^2}{E_T+eA_4}\right)^2} \tag{5.85}$$

$$V_{e,P} = c\gamma/(\hat{\gamma}_a B_e) \equiv c\frac{\gamma}{\hat{\gamma}}\frac{1}{\sqrt{1-\left(\frac{\gamma mc^2}{E_T+eA_4}\right)^2}}. \tag{5.86}$$

Furthermore, $\left(\frac{ds}{dt}\right)^2$ becomes

$$\left(\frac{ds}{dt}\right)^2 = \gamma^2 c^2 - \hat{\gamma}^2 {V_e}^2 . \tag{5.87}$$

The general expression (5.87) for $\left(\frac{ds}{dt}\right)^2$ due to a massive and electrically charged particle in a Schwarzschild field and an electrostatic field, can be written as

$$\left(\frac{ds}{dt}\right)^2 = \left(c\gamma^2 \frac{mc^2}{E_T + eA_4}\right)^2 . \tag{5.88}$$

V.8.4 Massless particles in a Schwarzschild field

Preliminarily note

Regarding the concept of the massless particle, for $\mathrm{m} = 0$, then

$$V_{m=0} = V_{P,m=0} = c\frac{\gamma}{\hat{\gamma}} . \tag{5.89}$$

In this case, we always have eqs. (5.72)-(5.75) and $\widehat{\boldsymbol{Vr}}$ is the angle made by the radius vector **r** and the velocity **V**. Considering eqs. (5.63) and (5.89), when $m = 0$, Eq. (5.76) becomes

$$\left(\frac{ds}{dt}\right)^2 = \gamma^2 c^2 - \hat{\gamma}^2 \left(c\frac{\gamma}{\hat{\gamma}}\right)^2 .$$

That is to say, for a massless particle in Schwarzschild potential

$$\left(\frac{ds}{dt}\right)^2 = 0 . \tag{5.90}$$

This agrees with Eq. (5.78) for $m = 0$. Taking here also the particle's trajectory in the plane $\theta = \pi/2$, then

$$V_{m=0}^2 = \dot{r}^2 + r^2\dot{\varphi}^2 . \tag{5.91}$$

Thus, if we consider (5.68), (5.69), and (5.86), we have

$$V_{m=0}^2 = \dot{r}^2 + r^2\dot{\varphi}^2 = V_{P,m=0}^2 = \left(c\frac{\gamma}{\hat{\gamma}}\right)^2 . \tag{5.92}$$

In addition, for $\dot{r}^2 = 0$ and the trajectory is a circle, since $\boldsymbol{rV_{m=0}} = 0$, then

$$\hat{\gamma} = 1 \tag{5.93}$$

and equation (5.91) becomes

$$V_{m=0} = V_{P,m=0} = r\dot{\varphi} = c\gamma \; . \tag{5.94}$$

CHAPTER VI

GENERALIZATION OF THE WAVES ASSOCIATED WITH PHOTONS TO WAVES ASSOCIATED WITH MASSIVE AND ELECTRICALLY CHARGED PARTICLES PROPAGATED IN THE ETHER

VI.1 Recalling the Elastic Interpretation of Electrodynamics

In Sec. IV, I considered the ether to be an elastic medium in which the field $\boldsymbol{\xi}$ of its point changes are governed by Eq. (3.5). Using the changes of the variables defined in (4.1)-(4.5), we can obtain the four equations (4.6)-(4.9), which *are identical to the Maxwell equations*. It follows that $\boldsymbol{E}$, $\boldsymbol{H}$, $\boldsymbol{B}$, $1/\eta$, $\boldsymbol{J}_e$, ρ_e, and ρ_0 represent the elastic natures of, respectively: the electrical field; the magnetic field; the magnetic induction; the inverse of the electric induction coefficient ε; the volumetric density of electrical currents; the volumetric density of electric charges; and the volumetric coefficient of the ether inertia. From (4.4) and (4.5), we can deduce that $\boldsymbol{J}_e$ and ρ_e verify the continuity equation (4.11) for electrical charges and currents, expressing charge conservation. I also demonstrated that the expressions for electromagnetic forces are the same as interactions of changes in the medium created by $\boldsymbol{C}$, i.e., by $e \equiv \rho_e(dL)^3$. For example, the force $\boldsymbol{f}_{vor}$ acting on a small element of the ether vortex $\boldsymbol{H}$, created by $e\boldsymbol{V}_e$, placed in an incident constant ether flow $\boldsymbol{B}$, is

$$\boldsymbol{f}_{vor} = e\boldsymbol{V}_e \wedge \boldsymbol{B} . \qquad (6.1)$$

This last equation is of the same form as that of the force acting on an electrical charge e, of velocity $\boldsymbol{V}_e$, in a magnetic induction $\boldsymbol{B}$. I also showed (ibid) that the electromagnetic field energy is the same as the energy of the medium changes created by $\boldsymbol{C}$, i.e., by e.

Since all the equations of the elastic interpretation of electrodynamics are identical to those of classical electrodynamics, it follows that these theories are identical, i.e., one can use any one of them. Moreover, one can affirm that the elastic interpretation of electrodynamics is the physical base of classical electrodynamics, as presumed by Maxwell and Einstein.

The Maxwell equations show that ρ_e and $\boldsymbol{J}_e$ create the relevant fields. Therefore, if we say that a particle possesses an electric charge e, we mean that e creates a Lienard-Wiechert field. Therefore, e can be considered an independent quantity that creates the field that interacts with the incident field, and this interaction is the force that the incident field exercises on e.

As such, I consider the elastic medium called the ether, submitted to a volumetric density of deforming force couples $\boldsymbol{C}$, which creates the field $\boldsymbol{\xi}$ of rotations on its points. They are governed by Eq. (3.8), called the "flat ether elasticity equation", which ensues from the Navier-Stokes-Durand equation of elasticity, and from which we can obtain the equations of electromagnetism.

In $D[0, Par(00)] \cap \Re(0,0)$, since

$$\frac{\rho_0}{\eta_0} \equiv \frac{1}{c^2}$$

equation (3.8) becomes

$$\boldsymbol{curl}(\boldsymbol{curl\xi}) = (\omega/c)^2\boldsymbol{\xi}\,. \tag{6.2}$$

Since, in the elastic interpretation of electrodynamics (EIE), we only use mechanical quantities, it follows that the dimensions used in EIE are MKS. In contrast, the dimensions used in classical electrodynamics (CE) are MKSA. From the facts that the EIE and the CE are identical theories, but in the EIE we use the MKS dimensions, while in the CE we use the MKSA dimensions, it follows that the MKSA dimensions are related to the MKS dimensions, as shown below.

If E_B denotes the magnetic part of the electromagnetic field energy, then we have, on one hand, $dim(E_B) = dim(\mu H^2 L^3)$, and on the other, $dim(E_B) = ML^2/T^2$. These two dimensional functions are identical, if one considers that $\boldsymbol{H} = \partial_t\boldsymbol{\xi}$, which implies that $dim(H) = L/T$ and $dim(\mu) = dim(\rho) = M/L^3$. Therefore, we can say that *the MKS dimensions of* $\boldsymbol{H}$*, of* μ*, and of* $\boldsymbol{B}$ *are respectively,* L/T*,* M/L^3*, and* $M/(TL^2)$. From (4.8) and $\dim(B) = M/(TL^2)$, we obtain $dim(\rho_e) = 1/L\,L$, that is, *the MKS* *dimension of* ρ_e is $1/L$. In the same way, it is easy to demonstrate that *the MKS dimensions of E are* M/LT^2.

VI.2 Waves Associated with Massive and Electrically Charged Particles in a Domain that is Void of their Generators

The generalization of the generator $G(0,0)$ of $Par(0,0)s$, defined in Sec. V.1, is the generator $G(m,e)$ of $Par(m,e\,)s$, and the generalization of $D[G(0,0), Par(0,0\,)s]$, or of $D[0, Par(0,0\,)s]$ is $D[G(m,e), Par(m,e\,)s]$, or $D[0, Par(m,e\,)s]$. The generator $G(m,e)$ will be addressed in a further publication, but, for the purposes of this unification, it is sufficient to consider only the physical phenomenon that occur in $D[0, Par(m,e\,)s]$. What one can say here is that it generalizes the density of couples of forces $\boldsymbol{C}$ applied to the ether.

The electromagnetic field $\boldsymbol{\xi}$, defined by Eq. (6.2), i.e., the field of rotations of ether points associated with $Par(00)$ in $D[0, Par(00)] \cap \Re(0,0)$ will now be generalized by the field $\boldsymbol{\xi}$ of rotations of ether points associated with $Par(m,e\,)$ in $D[0, Par(m,e)] \cap \Re(g,A)$, recalling that $\Re(g,A)$, denotes a spatial domain in which a gravitation field and an electromagnetic field exist. To that end, I demonstrate now the following proposition, but before this I consider the following precisions.

I denote $\boldsymbol{\xi}_{0\perp}$ that part of the vector amplitude $\boldsymbol{\xi_0}$ of the wave $\boldsymbol{\xi}$, which is perpendicular to the plane of the particle trajectory. This component $\boldsymbol{\xi}_{0\perp}$, contrary to the plane component of $\boldsymbol{\xi}_0$, can be taken as a constant. This is the case that I consider in the following. This is possible since, as shown above, we consider that the electron's trajectory in an atom is plane and that one of its amplitude components is perpendicular to the plane trajectory.

Proposition

The waves $\boldsymbol{\xi}$ defined in (5.21) or (5.22), i.e., for example

$$\boldsymbol{\xi} = \boldsymbol{\xi}_0 exp\left[i\omega\left(-t + \int \frac{dl}{V_P}\right)\right]$$

of $Par(m,e)s$ in $D[0, Par(m,e)]s \cap \Re(g,A)$, i.e., void of their generators, are solutions of the equation

$$\boldsymbol{curl}(\boldsymbol{curl\xi}) = \left[\frac{\omega^2}{{V_P}^2} - i\omega\partial_l\left(\frac{1}{V_P}\right)\right]\boldsymbol{\xi} \qquad (6.3)$$

which generalizes Eq. (6.2). Please note the following important remark.

Remark

In my previous publications, I did not always taken into account the term $i\omega\partial_l V_P$ because I considered it to be negligible, but it appears that, generally, it is necessary to take it into account as shown here.

Demonstration

Denoting

$$f \equiv exp(i\omega \int dl/V_P) \tag{6.4}$$

$$\beth' \equiv \left(\frac{\omega^2}{{V_P}^2} - i\omega\partial_l \frac{1}{V_P}\right) f\boldsymbol{e_2} \tag{6.5}$$

eqs (5.21) and (6.3) become

$$\xi = \xi_0 f exp(-i\omega t) \tag{6.6}$$

$$\boldsymbol{curl}(\boldsymbol{curl}\boldsymbol{\xi_0} f) = \beth'\xi_0\ . \tag{6.7}$$

Since $\boldsymbol{\xi}_0$ is constant and follows $\boldsymbol{e}_2$, (6.7) becomes

$$\boldsymbol{curl}[(\boldsymbol{grad}f)\wedge\boldsymbol{e}_2] = \beth'\boldsymbol{e}_2 \tag{6.8}$$

considering Eq. (6.4), and that

$$\boldsymbol{grad}f = i\omega f e_1/V_P \tag{6.9}$$

(6.8) becomes

$$i\omega\boldsymbol{curl}\left(\frac{f}{V_P}\boldsymbol{e}_1\wedge\boldsymbol{e}_2\right) = \beth'\ \boldsymbol{e}_2$$

that is,

$$i\omega\boldsymbol{curl}\left(\frac{f}{V_P}\boldsymbol{e}_3\right) = \beth'\ \boldsymbol{e}_2 \tag{6.10}$$

However,

$$\boldsymbol{curl}\left(\frac{f}{V_P}\boldsymbol{e}_3\right) \equiv \frac{f}{V_P}\boldsymbol{curle}_3 + \left[\boldsymbol{grad}\left(\frac{f}{V_P}\right)\right]\wedge\boldsymbol{e}_3$$

i.e.,

$$\boldsymbol{curl}\left(\frac{f}{V_P}\boldsymbol{e}_3\right) \equiv \left[\boldsymbol{grad}\left(\frac{f}{V_P}\right)\right]\wedge\boldsymbol{e}_3 \tag{6.11}$$

Therefore, considering (6.11), (6.10) becomes

$$i\omega\boldsymbol{grad}\left(\frac{f}{V_P}\right)\wedge\boldsymbol{e}_3 \equiv i\omega\left[\frac{\mathbf{1}}{V_P}\boldsymbol{grad}f\wedge\boldsymbol{e}_3 + f\boldsymbol{grad}\frac{\mathbf{1}}{V_P}\boldsymbol{e}_2\right] = ב'\boldsymbol{e}_2 \tag{6.12}$$

but

$$\left(\boldsymbol{grad}\frac{\mathbf{1}}{V_P}\right)\wedge\boldsymbol{e}_3 = \left(\partial_1\frac{\mathbf{1}}{V_P}\right)\boldsymbol{e}_1\wedge\boldsymbol{e}_3 = -\left(\partial_1\frac{\mathbf{1}}{V_P}\right)\boldsymbol{e}_2 \tag{6.13}$$

thus, (6.12) becomes

$$i\omega\boldsymbol{grad}\left(\frac{f}{V_P}\right)\wedge\boldsymbol{e}_3 \equiv i\omega\left[\frac{\mathbf{1}}{V_P}\boldsymbol{grad}f\wedge\boldsymbol{e}_3 - f\left(\partial_1\frac{\mathbf{1}}{V_P}\right)\boldsymbol{e}_2\right] = ב'\boldsymbol{e}_2$$

As is clear, $\boldsymbol{e}_1$ is the unit vector that follows the trajectory of such a $Par(m,e)$ around the nucleus and has to be denoted $\boldsymbol{e}_\ell$. As such, (6.13) has to be written

$$\boldsymbol{grad}\left(\frac{f}{V_P}\right)\wedge\boldsymbol{e}_3 = \left(\partial_\ell\frac{\mathbf{1}}{V_P}\right)\boldsymbol{e}_\ell\wedge\boldsymbol{e}_3 = -\left(\partial_\ell\frac{\mathbf{1}}{V_P}\right)\boldsymbol{e}_2$$

and (6.12) becomes

$$i\omega\boldsymbol{grad}\left(\frac{f}{V_P}\right)\wedge\boldsymbol{e}_3 \equiv i\omega\left[\frac{\mathbf{1}}{V_P}\boldsymbol{grad}f\wedge\boldsymbol{e}_3 - f\left(\partial_1\frac{\mathbf{1}}{V_P}\right)\boldsymbol{e}_2\right] = ב'\boldsymbol{e}_2\,. \tag{6.14}$$

Noting that $\boldsymbol{grad}f = i\omega f e_1/V_P$, introducing (6.9), in (6.14), we obtain

$$f\left[\frac{\mathbf{1}}{{V_P}^2}\omega^2 - i\omega\left(\partial_1\frac{\mathbf{1}}{V_P}\right)\right] = ב'$$

QED.
For the V_P constant, Eq. (6.3) becomes

$$\boldsymbol{curl}(\boldsymbol{curl\xi}) = (\omega/V_P)^2\boldsymbol{\xi}$$

proving that $\boldsymbol{\xi}$ defined in (5.21) or in (5.22) is a solution of Eq. (6.3) that generalizes Eq. (6.2). This is due to the fact that η_0 is defined in Eq. (6.1.1), i.e., $\rho_0/\eta_0 \equiv 1/c^2$, such that

$$\boldsymbol{curl}(\boldsymbol{curl\xi}) = \frac{\omega^2}{c^2}\boldsymbol{\xi}$$

becomes Eq. (6.3).

The real and imaginary forms of eqs. (6.2) and (6.3)

By separating the real and imaginary parts of Eq. (6.2), we obtain considering

$$\phi_P \equiv \omega(-t + x/c)$$

$$\boldsymbol{curl}(\boldsymbol{curl\xi_0} cos\phi_P) = \boldsymbol{\xi_0}\frac{\omega^2}{c^2}cos\phi_P \qquad (6.15)$$

and

$$\boldsymbol{curl}(\boldsymbol{curl\xi_0} sin\phi_P) = \boldsymbol{\xi_0}\frac{\omega^2}{c^2}sin\phi_P \qquad . \qquad (6.16)$$

Separating the real and imaginary parts of Eq. (6.3), and considering (5.42), we obtain the two following equations

$$\boldsymbol{curl}(\boldsymbol{curl\xi_0} cos\phi_P) = \boldsymbol{\xi_0}\frac{\omega^2}{V_P{}^2}\left(cos\phi_P - \frac{1}{\omega}sin\phi_P \cdot \partial_l V_P\right) \qquad (6.17)$$

and

$$\boldsymbol{curl}(\boldsymbol{curl\xi_0} sin\phi_P) = \boldsymbol{\xi_0}\frac{\omega^2}{V_P{}^2}\left(sin\phi_P + \frac{1}{\omega}cos\phi_P \cdot \partial_l V_P\right). \qquad (6.18)$$

We can see that eqs. (6.15) and (6.17) are the derivatives of eqs. (6.16) and (6.18) relative to Ω. Therefore, in order to analyze this state, it is sufficient to consider only, for example, eqs. (6.15) and (6.17), for which in this last equation, V_P is the phase velocity defined generally in Eq. (5.37). We can see that in the particular case where V_P is constant, for example on a straight line or on a circle, then Eq. (6.17) becomes

$$\boldsymbol{curl}(\boldsymbol{curl}\boldsymbol{\xi}_0 cos\phi_P) = \frac{\omega^2}{{V_P}^2}\boldsymbol{\xi}_0 cos\phi_P \ . \qquad (6.19)$$

In the particular case where $V_P = c$, then Eq. (6.17) becomes Eq. (6.15).

VI.3 More Explicit Forms of Equations (6.2) and (6.3)

Equation (6.3) can be written as follows

$$-\nabla^2\boldsymbol{\xi} + \boldsymbol{grad}(div\boldsymbol{\xi}) = \left[\frac{\omega^2}{{V_P}^2} - i\omega\partial_l\left(\frac{1}{V_P}\right)\right]\boldsymbol{\xi} \ . \qquad (6.20)$$

In our case, where $\boldsymbol{\xi}_0$ is constant and perpendicular to the trajectory, I show that we have the following relation

$$\boldsymbol{grad}(div\boldsymbol{\xi}) = 0 \ . \qquad (6.21)$$

Considering Eq. (5.22), that is, considering

$$\boldsymbol{\xi} = \boldsymbol{\xi}_0 e^{-i\omega t}e^{i\omega\int d\ell/V_P}$$

we have

$$\boldsymbol{grad}(div\boldsymbol{\xi}) = e^{-i\omega t}\boldsymbol{grad}\left[div\left(\boldsymbol{\xi}_0 e^{i\omega\int d\ell/V_P}\right)\right] \qquad (6.22)$$

However, $\boldsymbol{\xi}_0$ is constant and therefore (6.22) becomes

$$\boldsymbol{grad}(div\boldsymbol{\xi}) = e^{-i\omega t}\boldsymbol{grad}\left[\boldsymbol{\xi}_0 \cdot \mathbf{grad}\left(e^{i\omega\int d\ell/V_P}\right)\right] . \qquad (6.23)$$

The fact that $\boldsymbol{\xi}_0 \equiv \boldsymbol{e}_2\xi_0$ and $\mathbf{grad}\left(e^{i\omega\int d\ell/V_P}\right)$ follows the direction of $\boldsymbol{e}_\ell$ shows that

$$\boldsymbol{\xi}_0 \perp \mathbf{grad}\left(e^{i\omega\int d\ell/V_P}\right)$$

and therefore Eq. (6.21) is verified and Eq. (6.20) (i.e., Eq. (6.3)) can be written

$$-\nabla^2\boldsymbol{\xi} = \left[\frac{\omega^2}{{V_P}^2} - i\omega\partial_l\left(\frac{1}{V_P}\right)\right]\boldsymbol{\xi}\,. \tag{6.24}$$

As such, Eq. (6.17) can be generally written in the form

$$-\nabla^2(\boldsymbol{\xi_0}cos\Omega) = \frac{\omega^2}{{V_P}^2}\boldsymbol{\xi_0}\left(cos\Omega - \frac{1}{\omega}sin\Omega\;\partial_l V_P\right) \tag{6.25}$$

while eqs. (6.2) and (6.3) can be written, respectively

$$-\nabla^2\boldsymbol{\xi} = \frac{\omega^2}{c^2}\boldsymbol{\xi} \tag{6.26}$$

and

$$-\nabla^2(\boldsymbol{\xi_0}cos\Omega) = \frac{\omega^2}{c^2}\boldsymbol{\xi_0}cos\Omega\,. \tag{6.27}$$

The density of the ether is the constant ρ_0. This is confirmed by the fact that, as shown below, $\boldsymbol{\xi}$ *is constituted by immobile rotationally vibrating ether points.* The elasticity-restoring rotation coefficient η, defined in Eq. (1.25), which depends on the time and location in the ether, is related to the phase velocity and, together with relation (1.11), shows that n, as defined by

$$n \equiv \sqrt{\rho_0/\eta} = 1/V_P \tag{6.28}$$

is the refractive index in an anisotropic medium (Cf. Sec. 15.3.4 of Ref. 20). Furthermore, as shown in (6.5)

$\boldsymbol{\xi} \perp \boldsymbol{u}$, *i.e.,* $\boldsymbol{\xi}$ *is a rotating vibration of the ether.*

This accords well with the fact that, as shown above, the *density of the ether is constant*. Relation (6.28) is of the same form as for optical electromagnetism and this fact justifies Eq. (6.3) for the case in which the incident fields are time-independent.

Example

For the waves $\boldsymbol{\xi}$ of $Par(m,e)s$ in $\Re(g,0)$, we have

$$\boldsymbol{\xi}\cdot\boldsymbol{curl\xi} = i\omega\boldsymbol{\xi}\cdot(\boldsymbol{grad}S\wedge\boldsymbol{\xi}) = 0 \tag{6.29}$$

since $\boldsymbol{\xi} \perp \boldsymbol{gradS}\wedge\boldsymbol{\xi}$. For $Par(0,0)s$ in $\mathfrak{R}(g,0)_0$, ω becomes $2\pi\nu$, and (5.55) and (5.56) become

$$V_0 = V_{P0} = cg_{44}/G. \tag{6.30}$$

For $Par(m,0)s$ in $\mathfrak{R}(0,0)_0$, i.e., for free massive particles, we have

$$V_{P0} = c/B \tag{6.31}$$

$$V_0 = cB \tag{6.32}$$

where the expression for B is now

$$B = \sqrt{1-(mc^2/E_T)^2} \tag{6.33}$$

and that for $\boldsymbol{\xi}$, we now have

$$\boldsymbol{\xi} = \boldsymbol{\xi}_0 exp[\mathrm{i}\omega(-t + xB/c)] \,. \tag{6.34}$$

For $Par(0,0)s$ in $\mathfrak{R}(0,0)_0$, i.e., for the free electromagnetic wave, then

$$\boldsymbol{\xi} = \boldsymbol{\xi}_0 \exp\,[\mathrm{i}2\pi\nu(-t + \tfrac{x}{c})]$$

and we have

$$\boldsymbol{\xi} \perp \boldsymbol{e_x}$$

$$\boldsymbol{H} = |\boldsymbol{H}|\boldsymbol{e}_y$$

$$\boldsymbol{E} = |\boldsymbol{E}|\boldsymbol{e}_z \,.$$

VI.4 The Elastic Interpretation of the Fields that Act on a Particle

From Eq. (6.28), we can see that the elasticity-restoring rotation coefficient η is connected to the phase velocity V_P and the constant density ρ_0 of the ether by the relation $\eta = \rho_0 {V_P}^2$. Looking at (5.45), we can see that L_G, i.e., V_P defines the fields, by means of (6.3) and (6.28). As such, we can say that the fields are, finally, simply changes of the ether-restoring rotation coefficient η. In $\Re(g, A)$, in the presence of $Par(m, e)$, (5.37), and (5.45), we find that we cannot separate the gravitational field, the electromagnetic field, and the inertial mass m. In order to show the elastic side of the gravitational field, we consider $Par(0,0)$ in $\Re(g, 0)$, which, in the case (6.28), becomes

$$\eta = \eta_0 \left(\frac{g_{44}}{G - g_{j4}u^j}\right)^2 . \qquad (6.36)$$

This shows how the ether-restoring rotation coefficient η changes due to the gravitational field. For example, for $Par(0,0)$ in a Schwarzschild field, we have

$$\eta = \eta_0 (\gamma/\hat{\gamma})^2 \qquad (6.37)$$

where γ and $\hat{\gamma}$ are defined in (5.63). For $Par(m, 0)$ in $\Re(g, 0)$, we have

$$\eta = \eta_0 \left(\frac{g_{44}}{GB - g_{j4}u^j}\right)^2 . \qquad (6.38)$$

In a spatial region where there is an inertial mass point m, in a Schwarzschild field such that $m_0 \gg m$, then this m_0 together with m, changes η_0 into η. This is defined by

$$\eta = \eta_0 \left(\frac{\gamma}{\hat{\gamma}}\right)^2 \frac{{E_T}^2}{{E_T}^2 - \gamma^2 (mc^2)^2} . \qquad (6.39)$$

This shows how the restoring rotation coefficient η undergoes changes due to the gravitation field and to the inertial mass m of the massive particle subjected to this field. In the general case, η is given by two equations: (6.28) and (5.37).

CHAPTER VII

GENERALIZATION OF THE CONSTITUTION OF THE FREE PHOTON IN THE FRAME OF ETHER ELASTICITY TO THAT OF THE MASSIVE AND ELECTRICALLY-CHARGED PARTICLE

Equation (1.12) is here generalized and adapted to the general case where the particle is massive and subject to any field of forces. First, let us consider the solution of Eq. (1.12). Regarding this solution, we may recall that a photon is a "rotating ether trajectory segment" moving at the velocity of light. As I am going to demonstrate further, we note that the expression for these "free photon rotating ether trajectory segments", denoted $\boldsymbol{R}(\phi_P, \phi)$, is given in eqs. (1.15), (1.16), and (1.17).

Generalizing the constitution of the free photon in the frame of ether elasticity to the massive and electrically charged particle, I show that such a particle is also a specific sum $R(\phi_P, \phi)$ of waves $\boldsymbol{\xi}$. A "rotating ether trajectory segment" is thus formed that moves like $Par(m, e)$ and contains all of its parameters: in fact, a $Par(m, e)$ is a $R(\phi_P, \phi)$. It appears that the fields are also elastic changes in the ether. A rotating ether trajectory segment moves at velocity $\boldsymbol{V}$ along this trajectory. The expression for the angle of rotation denoted $\boldsymbol{R}(\phi_P, \phi)$ of this rotating ether trajectory segment is a function of ϕ, defined in (5.40), and of ϕ_P, defined in (5.41). This gives

$$\boldsymbol{R}(\phi_P, \phi) \equiv$$
$$\xi_0\, SINC[(\Delta\omega/2)\phi] \cdot \{[\mathbf{e}_z \wedge (\mathbf{V}/\mathrm{V})]\cos(\phi_P) + \mathbf{e}_z \sin(\phi_P)\}. \quad (7.1)$$

We can see that $\boldsymbol{R}(\phi_P, \phi)$ is a function of ℓ at a fixed instant t_0 and is a function of t at a fixed point ℓ_0. Let us now consider the component of this rotating ether trajectory segment $\boldsymbol{R}(\phi_P, \phi)$ in the plane (**V**, z), i.e., let us consider the expression for $\boldsymbol{R}(\phi_P, \phi) \cdot \mathbf{e}_z$. Considering (7.1), we have

$$\boldsymbol{R}(\phi_P,\phi)\mathbf{e}_z \equiv \xi_0\, SINC(\Delta\omega\phi/2)\cdot\sin(\phi_P) \tag{7.2}$$

$$[\boldsymbol{R}(\phi_P,\phi)\mathbf{e}_z]\mathbf{e}_z = \mathbf{e}_z\xi_0\, SINC(\Delta\omega\phi/2)\cdot\sin(\phi_P)\,. \tag{7.3}$$

Remark

For $\Delta\omega = 0$, (7.3) becomes $\mathbf{e}_z\xi_0\sin(\phi_P)$, which is a solution of Eq. 6.2) as discussed in Sec. IV.5.

I show now that $[\boldsymbol{R}(\phi_P,\phi)\mathbf{e}_z]\mathbf{e}_z$ is a specific sum of waves. For simplicity, I consider a $Par(m,e)$ in $\Re(g,0)_0$ and will use the following notations: if ω is fixed, and $\boldsymbol{\xi}$, ϕ, and ϕ' become functions of $\omega+\Delta\omega$, then they will be denoted $\boldsymbol{\xi}(\Delta\omega)$, $\phi(\Delta\omega)$, $\phi'(\Delta\omega)$, and if $\Delta\omega=0$, then they will be denoted simply $\boldsymbol{\xi}$, ϕ, and ϕ'. Following the increment theorem we can write, in this case where m is not zero

$$\phi(\Delta\omega) = \phi + \Delta\omega\phi' + (1{,}2)(\Delta\omega)^2\phi''(\theta\Delta\omega) \tag{7.4}$$

where $\phi''(\theta\Delta\omega)$ is the value of $\partial\phi'/\partial\omega$ when ω takes the value $\omega+\theta\Delta\omega$, ($0<\theta<1$)). Now, (5.44) and (5.55) yield

$$\phi''(\theta\Delta\omega) = -\frac{\hbar}{c}(mc^2)^2\int\sqrt{\frac{-u^iu^jg_{ij}g_{44}}{(E_T+\theta\hbar\Delta\nu)^2-g_{44}(mc^2)^2}}\,d\ell\,. \tag{7.5}$$

This equation shows that for a photon, $\phi''(\theta\Delta\omega)=0$, since $m = 0$. Let us now consider that the particle is an elementary massive one like, for example, a neutron. Since, in MKS unities, $\hbar = 10^{-34}$, and $m = 10^{-27}$, it follows that for the neutron, $(\hbar/c)(mc^2)^2 \cong 10^{-46}$. However, the value of the integral in (7.5) is finite, therefore, if the value of $\Delta\omega$ is also finite, then $\phi''(\theta\Delta\omega)$ is negligible in front of $\Delta\omega\phi$, except when $\phi=0$. However, this case is of no consequence since we consider $exp[i\phi(\Delta\omega)]$, as we will see. Thus, (7.4) can be written simply

$$\phi(\Delta\omega) = \phi + \Delta\omega\phi' \tag{7.6}$$

and for the massive and electrically charged particle, we can write

$$\boldsymbol{\xi}(\Delta\omega) = \boldsymbol{\xi}exp(i\Delta\omega\phi')\,. \tag{7.7}$$

Let us consider the following superposition $\hat{\boldsymbol{\xi}}(\Delta\omega)$ of waves $\boldsymbol{\xi}(\vartheta)$, namely

$$\hat{\boldsymbol{\xi}}(\Delta\omega) \equiv \frac{\mathbf{1}}{\Delta\omega}\int_{-\Delta\omega/2}^{\Delta\omega/2}\boldsymbol{\xi}(\vartheta)d\vartheta\ . \tag{7.8}$$

By virtue of (7.7), Eq. (7.8) becomes

$$\hat{\boldsymbol{\xi}}(\Delta\omega) \equiv \frac{\mathbf{1}}{\Delta\omega}\boldsymbol{\xi}\int_{-\Delta\omega/2}^{\Delta\omega/2} exp(i\vartheta\phi)d\vartheta \tag{7.9}$$

where ϕ is defined in eqs. (5.40) and (5.44), that is to say

$$\hat{\boldsymbol{\xi}}(\Delta\omega) \equiv \boldsymbol{\xi}_0\ SINC[(\Delta\omega/2)\phi]\exp\ (i\phi_P) \tag{7.10}$$

and where ϕ_P is defined in (5.42). Now, Eq. (7.10) can be written as

$$\hat{\boldsymbol{\xi}}(\Delta\omega) \equiv \boldsymbol{\xi}_0 SINC\left[\left(\frac{\Delta\omega}{2}\right)(\phi)\right](cos\phi_P + isin\phi_P)$$

that is also

$$-i\hat{\boldsymbol{\xi}}(\Delta\omega) \equiv [\boldsymbol{R}(\phi_P,\phi)\mathbf{e}_z]\mathbf{e}_z - \boldsymbol{i\xi}_0 SINC[(\Delta\omega/2)(\phi)]cos\phi_P\ . \tag{7.11}$$

We can see that

$$Real\ part\ of\ \left[-i\hat{\xi}(\Delta\omega)\right] = \ [R(\phi_P,\phi)e_z]e_z \tag{7.12}$$

for which the expression is given in Eq. (7.3). $[\boldsymbol{R}(\phi_P,\phi)\boldsymbol{e}_z]\boldsymbol{e}_z$ is the expression for the component in the fixed plane (x, z) of $\boldsymbol{R}(\phi_P,\phi)$ defined in Eq. (7.1). That is, Eq. (7.1) generalizes Eq. (7.10) and represents a massive and electrical charged particle submitted to gravitational and electrical fields. Such a particle is a rotating ether trajectory segment. At time t and at point ℓ of its trajectory, that is, for a given ϕ and ϕ_P, the expression for $\boldsymbol{R}(\phi_P,\phi)$ of this rotating ether trajectory segment, is given in Eq. (7.1). We can see that the maximal value of $\boldsymbol{R}(\phi_P,\phi)$ is that of $\boldsymbol{R}(0,0)$, defined by

$$\boldsymbol{R}(0,0) \equiv \xi_0[\mathbf{e}_z \wedge (\mathbf{V}/\mathrm{V})] \tag{7.13}$$

for which the absolute value is

$$|\boldsymbol{R}(0,0)| \equiv \xi_0\ . \tag{7.14}$$

We define now the particle dimensions, that is, its length and radius rotations. These radius rotations depend on their emplacement. The particle's length is defined from the point on the trajectory where the rotation is at its maximum, that is, from the point where ϕ is such that

$$SINC\left(\frac{\Delta\omega}{2}\phi\right) = 1 \qquad (7.15)$$

to the point on this trajectory where the particle rotation is considered to be very small. This is where

$$SINC\left(\frac{\Delta\omega}{2}\phi\right) \text{ is very small}$$

that is, where it takes the very small value ϵ when ℓ takes the value ℓ_ϵ, defined by

$$SIN\left(\frac{\Delta\omega}{2}\frac{\ell_\epsilon}{V_P}\right) \cong \frac{2V_P}{(\Delta\omega\ell_\epsilon)} \cong \epsilon\ . \qquad (7.16)$$

As such, the expression for the length denoted ℓ_ϵ of this bulb, that is, of this massive and electrical particle, is defined by

$$\ell_\epsilon \cong 2V_P/(\Delta\omega\epsilon)\ .$$

where ℓ_ϵ denotes the distance going from the point in the photon where the rotation is at its maximum, i.e., from the point $|\boldsymbol{R}(0,0)|$ to the point in the photon where the rotation is very small. This is where the free particle's rotation is $|\boldsymbol{R}(\phi_P,\phi)|$, defined by

$$|\boldsymbol{R}(\phi_P,\phi)| \cong \epsilon\xi_0\ .$$

$\boldsymbol{R}(\phi_P,\phi)$ defined in (7.1) and called the rotating ether trajectory segment represents a particle the rotation value of which is non zero only in the volume defined by $SINC\left(\frac{\Delta\omega}{2}\phi\right)$, which moves at velocity $\boldsymbol{V}$. Thus, the rotating ether trajectory segment $\boldsymbol{R}(\phi_P,\phi)$ describes all the physical properties of the massive and electrically charged particle, i.e., its trajectory, its velocity, its energy, its undulating mechanical nature, and its size.

CHAPTER VIII

THE ELASTO-GRAVITATIONAL EXPLANATION OF THE "STRONG INTERACTION"

VIII.1 Introduction

This chapter follows our previous investigations (Cf. also refs. 10, 13, and 11) on the elasto-gravitational theory and completes them. To recap, in these investigations I showed that:

a) Maxwell's electromagnetism emerges from the elasticity theory of a medium (Cf. also Ref. 13), being the ether, in which electrical charges are due to couples of forces applied to the medium, also creating the electromagnetic field. This fact is sufficient to affirm that the ether exists, as presumed by Maxwell and Einstein, without demonstration (Cf. quots. 1-3 in Sec. I.)

b) The ether elasticity theory implies, as shown above, that the electromagnetic waves are generalized by waves $\boldsymbol{\xi}$ associated with massive particles.

c) Particles and fields are due to changes in the ether (Cf. also refs. 10 and 13), formed in such a way that a particle is a $\boldsymbol{R}(\phi_P, \phi)$, formed by a superposition of waves, and agreeing with Einstein's opinion (Cf. Ref. 6):

….. the combination of the idea of a continuous field with the conception of a material point discontinuous in space appears inconsistent…..

It is known that, in the Newton-Maxwell case, the electrical repulsion between two massive electrically charged particles of the same sign are always greater than their Newtonian attraction. However, as I demonstrate below, by using the results of a)-c), i.e., by taking into account general relativity and the ether theory, their mutual gravitational attraction surpasses their electrical repulsion once they are sufficiently close one to the other.

Before presenting the mathematical demonstration of this phenomenon, I offer its qualitative explanation.

Let us consider two particles: one of mass m_0 and electric charge q_0, and the second of mass m and of electric charge q, such that m_0 is much greater than m, and immobile. Therefore, m_0 creates a Schwarzschild field, while q_0 creates the temporal component A_4 of the electromagnetic potential tensor. I show that this Schwarzschild field modifies the elasticity-restoring rotation coefficient η of the ether. The coefficient η is related to the phase velocity V_P of the wave $\boldsymbol{\xi}$ associated with the particle of mass m and electric charge q by the relation

$$\eta = \eta_0 {V_P}^2/c^2 \tag{6.28}$$

(Cf. Sec. VI and also Sec. VI of Ref. 13) η is also the inverse of the electric induction coefficient ε. Therefore, η, determined as the coefficient of A_4 in the expression for the Lagrange-Einstein function L_G, which of course also depends on the Einstein tensor $g_{\mu\nu}$. Looking at L_G, we can deduce that the distance r_0 exists, separating these two particles such that, for $r < r_0$, their mutual attractive gravitational force is greater than their mutual electrical repulsion. This is due to the fact that, at the Schwarzschild horizon, the attractive force is infinite and is very large near to it. In contrast, the Coulomb force is null at this horizon and very small as comes near to it due to the expression for η. This phenomenon plays the role of the "strong interaction" and explains the processes of fission and fusion.

A. Salam and J. Strathdee (Cf. Ref. 21) solved the Einstein equations for strong gravity in the limit at which weak gravity is neglected. The class of solutions they found can be reduced to the Schwarzschild solution with the weak-gravity Newtonian constant being replaced by a strong-coupling parameter.

The principal difference to our solution is that the parameter k, defined in Sec. V.8.2, must be a constant, as it follows from Einstein's gravitational field equation (Cf. the expression for α in those for g_{44} and g_{11} in Sec. V.8 above). As I show, it is gravity that becomes infinite at the Schwarzschild horizon and grows very large as one draws near to it. The Coulomb force vanishes once it is reached and is very small near to it. This is due to the fact that the elasticity-restoring rotation coefficient η (the inverse of the electric induction coefficient), is modified by the gravitation in such a way that it vanishes at the Schwarzschild horizon and is very small near it. These two phenomena, namely, the increasing gravitation near the Schwarzschild horizon and the diminishing Coulomb force there, results in the "strong interaction".

A. O. Barut (Cf. Ref. 22) has discussed the possibility that the strong interaction may be a consequence of the very short distance of the interaction of magnetic dipoles. Here, we discuss the electro-gravitational interaction in a somewhat different context and show that the gravitational interaction plays an important role as well.

VIII.2 Notation and Discussion of Some Results

A. Generalities

A particle of electric charge q_0 and mass m_0, such that $m_0 >> m$, m being the mass of a particle under the influence of the particle of mass m_0 considered to be immobile at the origin **O**, creates a Schwarzschild gravitational field and an electrostatic field. Since these fields are time-independent, the total energy E_T of the particle of mass m and charge q subject to these fields is constant (Cf. Sec. V.3). The quantities $g_{\mu\nu}$, γ^2, and $\hat{\gamma}^2$ are defined in Sec. V.8. The function L_G, denoted L_{GS}, (S for static) is then (Cf. Eq. (4) of Ref. 23)

$$L_{GS} = -mc\sqrt{c^2\gamma^2 - V^2\hat{\gamma}^2} + qA_4 \, . \qquad (8.1)$$

where $\boldsymbol{V}$ denotes the velocity of a particle of mass m and of electric charge q in a Schwarzschild gravitational field and in an electrostatic field, as well as A_4, the temporal component of the electromagnetic potential tensor, defined more precisely in the following. In order to determine the "strong interaction", we need to determine the expressions for $\boldsymbol{V}$ and A_4. This last requires the expression for the phase velocity $\boldsymbol{V}_P$ of the wave $\boldsymbol{\xi}$ associated with this particle.

B. Expression for the velocity V of an electrically charged particle subjected to a Schwarzschild gravitational field and an electrostatic field

From (5.7) and $P_4 = -E_T/c$, we can deduce

$$mc^2/ds = (E_T + qA_4)/dx_4 \qquad (8.2)$$

and from (5.7) and (8.2),

$$P_j = -(E_T + qA_4)dx_j/(\mathrm{c}dx_4) \, . \qquad (8.3)$$

Using the notation of (5.25), the expression for ds and (8.2) yield the expression for (5.26)

$$(g_{4\mu}g_{4\nu} - b^2 g_{\mu\nu})dx^\mu dx^\nu = 0\,. \qquad (8.4)$$

Let G and B denote the quantities defined by

$$G = \sqrt{-u^i u^j g_{ij}\gamma^2} \equiv \gamma\hat{\gamma} \qquad (8.5)$$

$$B = \sqrt{(1 - \gamma^2/b^2)}$$

by solving (8.4) for u^4 of a particle of mass m and electric charge q, we obtain

$$u^4 = G/(\gamma^2 B) \qquad (8.6)$$

and, considering that $u^4 \equiv cdt/d\ell \equiv c/V$, we also obtain

$$V = cB\gamma/\hat{\gamma}. \qquad \{68)$$

C. Expression for the phase velocity V_P of the classical wave ξ associated with this particle

In order to determine the expression for V_P, we can make use of direct reasoning and use the well-known relation (1.4). From this relation and (5.68), it follows that the expression for V_P is

$$V_P = c\gamma/(\hat{\gamma}B)\,. \qquad (5.69)$$

Let ϕ_P denote the quantity defined by

$$\mathrm{d}\,\phi_P = L_{GS}dt/\hbar \qquad (8.7)$$

which, considering (5.8) along a geodesic, yields (5.17). Since $\omega = E_T/h$ is an angular frequency, ϕ_P is the phase of a wave ξ associated with this particle. It is of the form (5.22) where ξ_0 is a constant vector perpendicular to the particle trajectory. The phase velocity V_P of ξ is defined by $\phi_P = 0$

on a ray when $d\ell = V_P dt$, i.e., when considering (5.17), and (5.20), (6.10), it takes the form

$$\xi = \xi_0 exp\left[i\omega\left(-t + \int \frac{d\ell}{V_P}\right)\right] \tag{8.8}$$

where the phase velocity V_P is given by (5.69).

D. The relation connecting V_P and the elasticity-restoring rotation coefficient η in a Schwarzschild gravitational field and an electrostatic field

Recalling that, since $c^2 = \eta_0/\rho_0$, the electromagnetic wave ξ, defined by Eq. (6.2), is associated with the free photon where ρ_0 denotes the constant volumetric density of the ether and η_0 denotes its constant elasticity-restoring rotation coefficient. As discussed in Chap. I, in the electromagnetic interpretation, $1/\eta_0 = \varepsilon_0$ appears to be the free electric induction coefficient and ρ_0 the free magnetic induction coefficient. We recall that the electromagnetic ξ, i.e., the field of rotations of the points of the ether associated with the photons, i.e., to $Par(0,0)$, is generalized by the field ξ of the rotations of the points of the ether associated with a particle of mass m and of electric charge e, i.e., a $Par(m, e)$, in a gravitational field and an electric field. Here, we consider that this gravitational field is a Schwarzschild field and the electric field is electrostatic. As I have demonstrated, a wave ξ defined by (8.8) and (5.69) is a solution of Eq. (6.3), where V_P, defined in (5.69), is related to the elasticity-restoring rotation coefficient η by the relation (1.25), i.e., Eq. (6.3) generalizes Eq. (6.2), and Eq. (6.28) generalizes Eq. $c^2 = \eta_0/\rho_0$.

However, we may think that (6.28) should be written with ρ instead of ρ_0, but I have proven that ρ is the constant ρ_0. The phase ϕ_P of ξ is defined in (8.8) and can be written in the form defined in eqs. (5.41), (5.42). Denoting S defined by

$$S = \int d\ell/V_P \ .$$

S is then the eikonal of ξ defined, for example, in Eq. (5.22). From Eq. (6.10) and from the identity (6.11) we can deduce the relation (6.12). As shown in Ref. 13, ω is, in our case, so large that (6.3) can be written as in (6.13). Using again the same identity, we have, with the same approximation, the relation (6.14). If $\boldsymbol{u}$ denotes the unit vector along the ray, then (6.9) implies Eq. (6.19). Thus, eqs. (6.29) and (6.30) yield the relation (6.16). Furthermore, considering (6.28) and (6.30), we can see that $\xi \perp \boldsymbol{u}$, i.e., $\boldsymbol{\xi}$ **is a transversal vibration of the ether**. This confirms that the density ρ of the ether is constant, i.e., ρ_0, and $1/V_P$ is generally the refractive index n in an anisotropic medium.

VIII.3 The Elasto-gravitational Interpretation of the "Strong Nuclear Interaction"

In the ether subject to a Schwarzschild gravitational field and an electrostatic field, created by the electric charge e_0, the expression for the component A_4 of the electromagnetic potential tensor, is

$$A_4 = -e_0/(4\pi\varepsilon r)\,. \qquad (8.9)$$

where A_4 is not solely created by e_0, but is also influenced by the Schwarzschild gravitational field. Indeed, η, defined by $\eta \equiv 1/\varepsilon$, is the elasticity-restoring rotation coefficient of the elastic medium, i.e., the ether. As shown previously, η is related to the phase velocity V_P of the wave $\boldsymbol{\xi}$ associated with a particle of mass m and of electric charge e in these fields by the relations (6.28) and (5.69). In contrast to the gravitational force, the electrostatic force to which this particle is subject does not depend on its proper mass m, furthermore, in the absence of gravitation η has to be η_0. From these two reasons, it follows that, as in (8.9), the expression for η is

$$\eta = 1/\varepsilon = \gamma^2/(\varepsilon_0\hat{\gamma}^2)\,.$$

Furthermore, in this interaction, we should consider that a = 0, and therefore

$$A_4 = -q_0\gamma^4/(4\pi\varepsilon_0 r)\,. \qquad (8.10)$$

Let us consider again that e is such that $ee_0 > 0$ and $m_0 \gg m$, i.e., that a particle of mass m_0 and of charge e_0 remains immobile even in the presence of a particle of mass m and electrical charge e. If the general relativity were not taken into account, then the Coulomb repulsion, due to e and e_0, would always be greater than the Newtonian attraction due to m and m_0, i.e., these two particles could not be bound. However, if general relativity and the elastic properties of the ether are taken into account, the gravitational attraction of these two particles can surpass their electrical repulsion. Indeed, inserting the expression (5.68) for the velocity V into (8.1), we obtain the following form of L_{GS}

$$L_{GS} = -\frac{(mc^2\gamma)^2}{E+qA_4} + eA_4 \tag{8.11}$$

which, considering again (8.1), for $V = 0$ becomes

$$L_{GS,V=0} = -mc^2\gamma + eA_4 \,. \tag{8.12}$$

This expression (8.11) shows that the sole coordinate upon which L_{GS} depends is r. It follows that the force $\boldsymbol{F}$, to which the particle of mass m and of charge q is subject, is radial. I demonstrate now the following proposition.

Proposition

The general relativistic gravitation and the elastic properties of the ether imply that there is a distance r_0 between a particle of mass m and of charge q and a particle of mass m_0 and of electric charge q_0, such that, for $r < r_0$, their attractive gravitational force is, even for $ee_0 > 0$, greater than their electric repulsion. This phenomenon plays the role of the "strong nuclear interaction".

Demonstration

A particle of mass m and of electric charge e is subject to the radial force

$$F(r) = \partial_r L_{GS} \tag{8.13}$$

for which, considering now (8.1) and (8.10), the expression is

$$F(r) = -\frac{mc[c^2\alpha/r^2 - \partial_r(V\hat{\gamma})^2]}{2\sqrt{c^2\gamma^2 - V^2\hat{\gamma}^2}} + \frac{ee_0\gamma^2}{4\pi\varepsilon_0 r^2}\left(1 - \frac{3\alpha}{r}\right). \tag{8.14}$$

Considering Eq. (43) of Ref. 10, Eq. (8.13), and $\alpha = 2m_0k/c^2$, for $V = 0$, we have

$$F(r) = \frac{1}{r^2}\left[-\frac{mm_0k}{\gamma} + \frac{ee_0\gamma^2}{4\pi\varepsilon_0}\left(1 - \frac{3\alpha}{r}\right)\right]. \tag{8.15}$$

This equation shows that, firstly

$$F(\alpha) = -\infty \tag{8.16}$$

and $\gamma = 0$, that is, $Par(m, e)$ is then subject to a very strong attractive force from the nucleus. For

$$0 < mm_0 k < ee_0/(4\pi\varepsilon_0) \tag{8.17}$$

which is generally the case, a value r_0 of r exists that is different from α, such that

$$F(r_0) = 0 \tag{8.18}$$

which is where

$$\frac{4\pi\varepsilon_0 mm_0 k}{ee_0} = \left(1 - \frac{\alpha}{r_0}\right)^{3/2} \left(1 - \frac{3\alpha}{r_0}\right). \tag{8.19}$$

Indeed, considering (8.17), this solution r_0 exists and is greater than 3α, the exact one being obtainable through numerical calculation and e can be immobile at r_0.

I consider now $F(R)$ for $R >> r_0$, and for $V \neq 0$. In this case, L_G is denoted $L_{G,R}$ and Eq. (8.1) becomes

$$\sqrt{c^2 - V^2} - \frac{ee_0}{4\pi R} \tag{8.20}$$

that is

$$F(R) \cong \frac{ee_0}{4\pi} \frac{1}{R^2} \tag{8.21}$$

$Par(m, e)$, located at R, is then subject to a repulsive force.

Finally, in summary:

$r \in [\alpha, r_0[\ \Rightarrow F(r) \leq 0$, i.e., $Par(m, e)$ and $Par(m_0,\ e_0)$ mutually attract each other, this is the so-called strong interaction;
$r = r_0\ \Rightarrow\ F(r_0) = 0$, i.e., $Par(m, e)$ is free at r_0;
$r \in]r_0, \infty[\ F(r) > 0$, i.e., $Par(m, e)$ and $Par(m_0, e_0)$ mutually repulse each other;
$r = \infty\ \Rightarrow\ F(\infty) = 0$, i.e., $Par(m, e)$ is free and moves at a constant velocity.

Remark

It appears that the external particle that collides with the $Par(m, e)$ and causes $r > r_0$, causes its ejection. But, if this collision causes $r < r_0$, then this collision causes to this $Par(m, e)$ to be bound. This is a fusion process called "strong interaction".

CHAPTER IX

QUANTUM MECHANICS AND SOME OTHER RESULTS SHOWN TO ENSUE FROM ETHER ELASTICITY

IX.1 Introduction

The Schrödinger equation comes from the following axiom:

> In Cartesian coordinates and in the absence of gravitation, the operator $-i\hbar\eth_\mu$ is associated with the component P_μ of the momentum tensor.

This axiom, inspired from the de Broglie plane wave, is **arbitrary**, and, therefore, so is quantum mechanics, even though it yields very important results. In Ref. 24, I showed that quantum mechanics may be extended to the case in which a gravitational field is present by replacing the operator $-i\hbar\eth_\mu$ by the covariant operator $-i\hbar D_\mu$, but this last axiom is also arbitrary. As such, we may think that the theory of quantum mechanics can be generalized as a theory based on the concepts of time, the ether, forces, movements, and relativity, and not just on axioms.

I show below that quantum mechanics appears to be a particular case of the "ether elasticity theory", and is also compatible with general relativity (Cf. also refs. 10, 13, and 15). Regarding the ether, we should recall that both Maxwell and Einstein presumed its existence. Indeed, Maxwell wrote (Art. 866 of Ref. 1 and rewritten in Sec. I.1): "*Hence all these theories lead to the conception of a medium.....*". Einstein (Ref. 3. rewritten (ibid)) stated: "*Recapitulating, we may say that according to the general theory of relativity, space is endowed with physical qualities; in this sense, therefore, there exists an ether…….*".

As shown here and in previous papers, the "ether elasticity equation" that governs the rotations $\boldsymbol{\xi}$ of ether points follows from the Navier-Stokes-Durand Eq. (3.1) of elasticity, taking into account, not only the densities of forces applied to the elastic medium, but also the densities of couples applied to it. It was E. Durand (Cf. Ref. 12) that completed the Navier-

Stokes equation by introducing these densities of couples. As proven here, we have to consider only the particular case where the ether is subjected to force couple densities, that is, where the ether is governed by the ether elasticity equations (30)-(32) of Ref. 13 and by, for example, Eq. (6.3), in which the point rotations $\boldsymbol{\xi}$ are waves of constant frequency. In this case, the ether elasticity theory generalizes the quantum mechanics since it yields the Schrodinger equation. It does not just yield the Schrodinger equation, for which the solutions are scalar state functions, but an equation for which the solution is vectorial waves $\boldsymbol{\xi}$, called "particle waves". This new equation that follows from the Navier-Stokes-Durand equation is valid even in the presence of a gravitational field.

These waves $\boldsymbol{\xi}$, called "particle waves" because they are associated with $Par(m,e)s$, (i.e., particles of mass m and electric charge e), can be described as follows: an adequate superposition $\hat{\boldsymbol{\xi}}(\Delta\omega)$ of waves $\boldsymbol{\xi}$ forms $\boldsymbol{R}(\phi_P,\phi)$ that moves like a $Par(m,e)$ and therefore a $Par(m,e)$ can be considered to be a $\boldsymbol{R}(\phi_P,\phi)$. In particular, a $Par(0,0)$ is a photon and the waves $\boldsymbol{\xi}$ associated with it are electromagnetic.

A $Par(m,e)$ in a bound state subjected to, for example, a Schwarzschild field and a Coulomb field, due to a $Par(m_0,e_0)$ being immobile at origin, the waves $\boldsymbol{\xi}$ that compose $\hat{\boldsymbol{\xi}}(\Delta\omega)$ interfere with themselves, since they are located around the nucleus. Due to this interference, only certain waves $\boldsymbol{\xi}$ remain that do not self-destruct, but on the contrary, are amplified, i.e., they resonate. This resonance happens when a closed trajectory of a $Par(m,e)$ contains an integer number "n" of wave lengths I_w, and these resonant waves are equivalent to the so called "quantum state" of quantum mechanics.

As such, the ether elasticity theory presented here yields more results than the Schrodinger equation on its own, since it relies on the physical properties of a specific elastic medium the ether in which waves propagate and for which the fields are changes in its elastic properties. Finally, both the ether elasticity theory and the theory of quantum mechanics yield the Schrodinger equation, but this equation is only a particular case of the ether elasticity theory, as we now go on to prove.

IX.2 Recalling the Equations that Govern the Ether and their Solutions

We consider that the fields to which a $Par(m,e)$ is subjected are static. In this case, (Cf. Sec. VI) the particle waves $\boldsymbol{\xi}$ associated with $Par(m,e)$ are the solutions of the ether elasticity wave equation (6.2), where $\boldsymbol{V}_P$ denotes the constant phase velocity of $\boldsymbol{\xi}$ and ω denotes a constant pulsation. As shown, $\boldsymbol{\xi}$ is of the form given in (5.21) or (5.22) in which the phase ϕ_P is related to the Lagrange-Einstein function, L_G, of a $Par(m,e)$ in $\Re(g,A)$, the expression for which is defined in Eq. (5.1) by the relation (5.15) and where $\boldsymbol{\xi}_0$ denotes a vector that depends on the spatial coordinates x^i of $Par(m,e)$. The phase ϕ_P follows from L_G as following. Since the expression for the momentum tensor P_μ is defined by

$$P_\mu \equiv \partial L_G/\partial \dot{x}^\mu \tag{9.1}$$

is given in (5.7). Considering (5.8), it follows that

$$L_G dt = P_\mu x^\mu \equiv P_4 x^4 + P_i x^i \equiv -E_T dt + P_i x^i \,. \tag{9.2}$$

V_P is defined by Eq. (5.20) and it follows that for a $Par(m,e)s$ subjected to static fields we have

$$\phi_P = \omega(-t + \textstyle\int d\ell/V_P) \tag{9.3}$$

where the spatial curvilinear integral is taken along a trajectory defined in Sec. V.7 and where $\boldsymbol{V}_P$, for which the expression is given below for the cases that interest us, is the phase velocity of a wave $\boldsymbol{\xi}$, while ω denotes the pulsation defined by $\omega \equiv E_T/\hbar$. In this equation, E_T denotes the total energy of $Par(m,e)$. This energy is constant since the fields to which it is subject are static. Furthermore. E_T can be written in the form

$$E_T = mc^2 + h\nu \tag{9.4}$$

where ν is a constant frequency. As shown (ibid) for a $Par(m,e)$, an electron $\boldsymbol{\xi}_0$ is perpendicular to the trajectory. The velocity V of $Par(m,e)$ is related to V_P by the relation (1.4) and, therefore, if we define ϕ by (4.43), we obtain the relation (5.40).

IX.3 The Schrodinger Equation as a Particular Form of the Ether Elasticity Equation

When a $Par(m,e)$ is subjected to a Schwarzschild field and a Coulomb field created by a $Par(m_0,q_0)$, i.e., a particle of mass m_0 and electric charge e_0 immobile at the point of origin O, then, considering Eq. (5.37) and eqs. (5.61)-(5.64), the expression for V_P is

$$V_P{}^2 = \frac{c^2 E_T{}^2 \gamma^2}{\hat{\gamma}^2[(E_T+eA_4)^2-\gamma^2(mc^2)^2]} \tag{9.5}$$

where

$$eA_4 \equiv -|ee_0|/(4\pi\varepsilon r)\ . \tag{9.6}$$

Furthermore, from the identity

$$(E_T + eA_4)^2 - \gamma^2(mc^2)^2 \equiv (2mc^2 + h\nu + eA_4)(h\nu + eA_4) + (mc^2)^2\alpha/r \tag{9.7}$$

it appears that, in the "non-relativistic approximation", i.e., when α/r is neglected in front of 1 and $h\nu + eA_4$ in front of mc^2, but where $(mc^2)^2\alpha/r$ is not neglected, Eq. (9.7) becomes

$$(E_T + eA_4)^2 - \gamma^2(mc^2)^2 \cong 2mc^2(h\nu + \hat{a}/\mathrm{r}) \tag{9.8}$$

where $\hat{a}$ is defined by

$$\hat{a} \equiv \frac{|ee_0|}{(4\pi\varepsilon)} + mm_0k \tag{9.9}$$

and the expression for the phase velocity (9.5) becomes then

$$V_P{}^2 = \frac{E_T{}^2}{2m(h\nu+\hat{a}/\mathrm{r})}\ . \tag{9.10}$$

Denoting by $\boldsymbol{\Psi}$ the vector defined by

$$\boldsymbol{\Psi} \equiv \boldsymbol{\xi}_0 exp(i\omega \int d\ell/V_P) \tag{9.11}$$

and considering that $\boldsymbol{\xi} = \boldsymbol{\xi}_0 exp(i\phi)$ can be written

$$\boldsymbol{\xi} = \boldsymbol{\Psi} exp(-i\omega t) \tag{9.12}$$

by inserting (9.12) in (6.3) for a circular trajectory, we obtain

$$\boldsymbol{curl}\left(V_P{}^2\boldsymbol{curl\Psi}\right) = \omega^2\boldsymbol{\Psi}\ . \tag{9.13}$$

Using the following two identities

$$\boldsymbol{curl}(b\boldsymbol{B}) \equiv b\boldsymbol{curlB} + \boldsymbol{grad}b \wedge \boldsymbol{B}$$

and

$$\boldsymbol{curl}(\boldsymbol{curlB}) \equiv -\nabla^2\boldsymbol{B} + \boldsymbol{grad}(div\boldsymbol{B}) \tag{9.14}$$

where b denotes any scalar and $\boldsymbol{B}$ any vector, Eq. (9.13) takes the form

$$-\nabla^2\boldsymbol{\Psi} + \Xi = \omega^2\boldsymbol{\Psi}/{V_P}^2 \tag{9.15}$$

where Ξ is defined by

$$\Xi \equiv \left[\boldsymbol{grad}(logV_P{}^2)\right] \wedge \boldsymbol{curl\Psi}\,. \tag{9.16}$$

On a circular_trajectory, ${V_P}^2$ is constant and $\Xi = 0$, that is, considering the expression for ${V_P}^2$ for $Par(m, e)$ subjected to the field created by $Par(m_0, q_0)$, then, on account of (9.5), Eq. (9.15) becomes

$$-\nabla^2\boldsymbol{\Psi} = \left(\frac{\hat{\gamma}}{c\gamma\hbar}\right)^2 [(E_T + eA_4)^2 - \gamma^2(mc^2)^2]\boldsymbol{\Psi}\,. \tag{9.17}$$

However, in the non-relativistic approximation, on account of (9.10), Eq. (9.17) becomes

$$-\frac{\hbar^2}{2m}\nabla^2\boldsymbol{\Psi} = (h\nu + \hat{a})\boldsymbol{\Psi} \tag{9.18}$$

which, considering (9.12), can be written

$$-\frac{\hbar^2}{2m}\nabla^2\xi = i\hbar\eth_t\xi + \hat{a}\xi\,. \tag{9.19}$$

Equation (9.19) is Schrodinger's equation and I have therefore demonstrated that this equation is a particular form of the ether elasticity equation (6.3), i.e., that:

the ether elasticity theory generalizes the quantum mechanics.

IX.4 Some Other known Results from the Ether Elasticity Theory

A. The Bohr-Sommerfeld condition

For simplicity's sake, I consider a $Par(m,0)$ in $\Re(g,0)_0$ and use the following notations. If ω is fixed and ξ, ϕ_P, and ϕ are functions of $\omega+\Delta\omega$, then they are denoted $\xi(\Delta\omega)$, $\phi_P(\Delta\omega)$, and $\phi(\Delta\omega)$. If $\Delta\omega=0$, then they are simply denoted ξ, ϕ_P, and ϕ. As I have shown in Sec. VII.2 with very good approximation, we have the relations (7.6)-(7.11).

With a $Par(m,e)$ in a bound state subjected to a Schwarzschild and a Coulomb field, due to a $Par(m_0,e_0)$, the waves ξ that compose $\hat{\xi}(\Delta\omega)$ interfere with themselves. However, a certain wave ξ does not destroy itself through this interference, but, on the contrary, resonates. This resonance happens when a close trajectory of a $Par(m,e)$ contains an integer number "n" of wave length I_w.

Let us consider the case where the trajectories described by a $Par(m,e)$ attracted by a $Par(m_0,e_0)$ are circular and located in the plane $\theta=\pi/2$. In such a circular state, I_w is constant since it does not depend on the angle φ described by the constant radius vector r, denoted then r_c. The resonance condition is such that

$$2r_c/I_w\,. \tag{9.20}$$

The expression for the total frequency ν_T is

$$\nu_T=E_T/h \tag{9.21}$$

and since

$$I_w=V_P/\nu_T$$

the expression for I_w is

$$I_w=hV_P/E_T. \tag{9.22}$$

In this circular case, we can express V_P as a function of the component P_φ of the momentum tensor defined by

$$P_\varphi=\partial L_{Ge}/\partial\dot{\varphi} \tag{9.23}$$

where L_{Ge} is the Lagrange-Einstein function. In this case where the trajectories are circumferences of radius r_c in the plane $\theta = \pi/2$, the expression for L_{Ge} is

$$L_{Ge} = -mc\dot{s}_c + eA_4\dot{x}^4/\mathrm{c} \qquad (9.24)$$

where

$$\dot{s}_c = \sqrt{\gamma^2(\dot{x}^4)^2 - \dot{\varphi}^2 r_c{}^2}\,. \qquad (9.25)$$

From

$$\xi = \xi_0 exp(i\phi)$$

and (9.3), we can see that

$$V_P = E_T/P_i u^i$$

and since r_c is constant, it follows that

$$P_i u^i = \frac{P_\varphi d\varphi}{(r_c d\varphi)} = \frac{P_\varphi}{r_c}\,.$$

Therefore, V_P can be written in the following form

$$V_P = E_T r_c/P_\varphi \qquad (9.26)$$

and I_w as the following

$$I_w = hr_c/P_\varphi\,. \qquad (9.27)$$

Inserting this expression for I_w in (9.20), we retrieve the Bohr-Sommerfeld relation

$$P_\varphi = n\hbar \qquad (9.28)$$

which is a resonance condition for the wave $\xi = \xi_0 exp(i\phi_P)$ in a circular bound state.

B. The permitted eigenvalues of r and hν

I determine now the values of ν and r_c that cause this resonance by determining two equations involving them, from (9.23)-(9.25), it follows that, in this circular bound state

$$P_\varphi = mc\dot{\varphi}{r_c}^2/\dot{s}_c \equiv mcr_cV/\dot{s}_c \,. \tag{9.29}$$

Now, in space-time, the expression for velocity V in this circular motion is (Cf. refs. 10 and 13)

$$V = c\gamma\sqrt{1-\left(\frac{\gamma mc^2}{mc^2+h\nu+eA_4}\right)^2} \tag{9.30}$$

which, in the non-relativistic approximation and considering (9.8), becomes

$$V \cong \sqrt{\frac{2}{m}\left(h\nu+\frac{\hat{a}}{r_c}\right)} \,. \tag{9.31}$$

By inserting (9.31), in (9.29), It follows that

$$P_\varphi \cong r_c\sqrt{2m\left(h\nu+\frac{\hat{a}}{r_c}\right)} \tag{9.32}$$

i.e., considering (9.28), we obtain

$$(n\hbar)^2 \cong 2m{r_c}^2\left(h\nu+\frac{\hat{a}}{r_c}\right) . \tag{9.33}$$

Equation (9.33) connects ν and r_c, and, on account of (9.31), yields

$$V \cong n\hbar/(r_cm) \,. \tag{9.34}$$

In order to determine another equation of ν and/or of r_c, we can use the following motion equation

$$\frac{d}{dt}\frac{\partial L_{Gc}}{\partial \dot{r}} - \frac{\partial L_{Gc}}{\partial r} = 0 \,. \tag{9.35}$$

From (9.24), we can see that

$$\frac{\partial L_{Gc}}{\partial \dot{r}} = 0 \text{ , it follows also that } \frac{\partial L_{Gc}}{\partial r} = 0$$

and, considering (9.24), (9.35) becomes

$$-\frac{1}{2} mc \frac{\alpha c^2/r_c{}^2 = 2\dot{\varphi}^2 r_c}{\sqrt{\gamma^2 c^2 - V^2}} + e\partial_{r_c} A_4 = 0 \text{ .} \qquad (9.36)$$

Keeping in mind that $\dot{\varphi} r_c = V$ and $\partial_{r_c} A_4 = q_0/(4\pi\varepsilon_0 r_c{}^2)$, in the non-relativistic approximation, (9.36) becomes

$$m r_c V^2 - \hat{a} \cong 0 \qquad (9.37)$$

and, taking into account (9.34) and (9.9), we obtain

$$r_c \cong (n\hbar)^2/(\hat{a} m) \text{ .} \qquad (9.38)$$

Inserting (9.38) into (9.33), we obtain

$$h\nu \cong -\frac{\hat{a}^2 m}{2(n\hbar)^2} \qquad (9.39)$$

since $\lfloor e e_0 \rfloor/(4\pi\varepsilon) \gg m m_0 k$, Eq. (9.39) can also be written with a good approximation

$$h\nu \cong -\frac{m}{2(n\hbar)^2}\left(\frac{e e_0}{4\pi\varepsilon}\right)^2 \text{ .} \qquad (9.40)$$

We can see that the relations (9.40) and (9.38), obtained from the theory of the waves $\boldsymbol{\xi}$ in the elastic ether are the same as those obtained from the current theory of quantum mechanics. As such, it follows that there is a relation between these two theories and, moreover, that the ether elasticity theory generalizes the usual theory of quantum mechanics.

IX.5 Further Results Regarding the Equations Governing the Ether

In previous publications, I proved that the ether is governed by the Navier-Stokes-Durand equation of elasticity (Cf. Ref. 12); that changes to the ether, denoted $\boldsymbol{\xi}$, are rotations of spatially immobile ether points forming particles and fields (Cf., e.g., refs. 10 and 13); and that a monochromatic wave $\boldsymbol{\xi}$ of phase velocity $\boldsymbol{V}_P$ and of frequency ν, propagating in the ether, is the solution of the wave equation (6.3). On circularly trajectories, e.g., around an atomic nucleus, Eq. (6.3) becomes (9.13), the solutions of which are of the form given in Sec. V.4.

The sum of the waves $\boldsymbol{\xi}$ for ω being replaced by the interval $\left[\omega - \frac{\Delta\omega}{2}, \omega + \frac{\Delta\omega}{2}\right]$ forms a photon rotating ether trajectory segment $\boldsymbol{R}(\phi_P, \phi)$, denoted $\hat{\boldsymbol{\xi}}$, and defined in Sec. VII. As defined in the frame of the ether, $\hat{\boldsymbol{\xi}}$ denotes a particle moving at the group velocity $\boldsymbol{V}$ connected to $\boldsymbol{V}_P$ by the relation (1.4). Considering the resonant case defined by the fact that an electron subjected to nuclear attraction moves on a circle denoted $\hat{C}(r)$ of radius r. Thanks to destructive interference, only one $\boldsymbol{\xi}$ remains from the sum $\hat{\boldsymbol{\xi}}$, i.e., for

$$SINC[(\Delta\omega/2)(t - \int d\ell/V)] = 1 .$$

This remaining $\boldsymbol{\xi}$ superposes itself on $\hat{C}(r)$, propagates at the phase velocity V_{Pc}, and is of constant amplitude ξ_0. If ℓ denotes the length of the electron-wave on $\hat{C}(r)$, then ℓ is defined by

$$\int d\ell/V_{Pc} = (\theta + 2\pi K)\, r/V_{Pc} \qquad (9.41)$$

that is, $\ell = r(\theta + 2\pi K)$, where K denotes an integer, $\theta + 2\pi K$ is the total angle occupied by this electron-wave, and $\theta \in]0, 2\pi\,]$. Furthermore, the expression for V_{Pc} is given by eqs. (9.5) and (9.6), where e and e_0 are the electron and proton electric charges; A_4 is the electromagnetic potential due to e_0, which is subjected to e; and m is the electron mass. For this resonant case, let n_0 denote a positive integer such that $\hat{C}(r)$ contains n_0 wave lengths λ, that is

$$r\omega/V_{Pc} \equiv 2\pi r/\lambda = n_0 \qquad (9.42)$$

i.e.,

$$\omega \int d\ell/V_{Pc} = (\theta + 2K\pi)n_0 . \qquad (9.43)$$

In this case, the phase-wave associated with this electron, which is of significant length, wraps around the nucleus. The expression for the resonant phase-wave, denoted $\boldsymbol{\xi}(K, n_0)$ in this case instead of $\boldsymbol{\xi}$, is

$$\boldsymbol{\xi}(K, n_0) = \boldsymbol{\xi}_0 \Sigma_{k=0}^{k=K} sin[\omega t - (\theta + 2k\pi)n_0] \qquad (9.44)$$

that is,

$$\xi(K, n_0) = (1 + K)\xi_0 sin[(\omega t - \theta n_0)] \qquad (9.45)$$

which, for $\theta = (\omega t - \pi/2)/n_0$, becomes $(1 + K)\xi_0$, and for $\theta = \omega t/n_0$ becomes $zero$. Since from the general expression (7.7), the sole wave that remains in this atom layer is $\xi(K, n_0)$, it follows that $\xi(\Delta\omega)$ becomes $\xi(K, n_0)$, which is the expression for the resonant electron single-particle wave.

CHAPTER X

ELECTRON SPIN PROVEN TO BE A PARTICULAR CASE OF ETHER ELASTICITY

X.1 Introduction

Maxwell and Einstein presumed the existence of the ether (Cf. citations in Sec. I). In this context, I showed in Chap. II that the Maxwell equations of electromagnetism follow from the case of elasticity theory in which the elastic medium is the ether on which a volumetric density $\boldsymbol{C}$ of couples of forces that create the field $\boldsymbol{\xi}$ of the rotations of ether points acts. This field $\boldsymbol{\xi}$ is governed by the "ether elasticity equation", which follows from the Navier-Stokes-Durand equation and generates the Maxwell equations of general expression presented in Sec. III.

These results motivated me to investigate the common physical nature of the fields and of the particles that reduce the contradictions and inconsistency cited by Einstein (Cf. Sec. VII.3). To this end, it was natural to extend this elastic interpretation to the case where particles are not only photons, but can also be $Par(m,e)s$, i.e., particles of mass m and electrical charge e, which can be subjected to electromagnetic and/or gravitational fields. This extension has been achieved by showing that the Lagrange-Einstein function L_G of $Par(m,e)$ does not only yield the four motion equations of the particle, but also is such that ϕ_P, defined by $\hbar d\phi_P/dt = L_G$, is the phase of a wave associated with these particles. Thus, L_G implies the existence of waves, of phase ϕ, denoted $\boldsymbol{\xi}_{m,e}$, associated with these particles, that generalize the wave $\boldsymbol{\xi}_{0,0}$. In the absence of confusion all these waves will simply be denoted $\boldsymbol{\xi}$.

Let us consider $\boldsymbol{\xi}_{m,e}$ more deeply. Since, in general, waves are propagated in an elastic medium, it follows that, like the waves $\boldsymbol{\xi}_{0,0}$, the waves $\boldsymbol{\xi}_{m,e}$ are propagated in the ether. Because when m tends towards zero, $\boldsymbol{\xi}_{m,e}$ becomes, $\boldsymbol{\xi}_{0,0}$, which is a field of rotations of points of the ether, it follows that $\boldsymbol{\xi}_{m,e}$ is, like $\boldsymbol{\xi}_{0,0}$, a field of rotations of points of the ether. Since the waves $\boldsymbol{\xi}_{0,0}$ are the solution of the flat ether elasticity Eq. (3.5), it follows that the waves $\xi_{m,e}$ are solutions of an equation that generalizes this equation. Therefore, the particle trajectory equation, obtained from Einstein's particle motion equation, generalizes the light ray equation. Furthermore, it is important to note, as shown in previous chapters, that:

> A specific sum of $\xi_{m,e}$ waves forms a $\hat{\xi}(\Delta\omega)_{m,e}$, which moves like a $Par(m,e)$ and contains all its parameters.

In this chapter, we are interested *only in the fact that a* $Par(m,e)$ *is a* $\hat{\xi}_{m,e}$. In its motion, this $Par(m,e)$, i.e., this $\hat{\xi}(\Delta\omega)_{m,e}$, creates a Lienard-Wiechert covariant potential tensor from which we can deduce the electromagnetic field and, in particular, the magnetic field $\boldsymbol{H}$.

In the ether elasticity theory, *this* $\boldsymbol{H}$ *is the rotational velocity* $\partial_t\xi$ *of the ether points* (Cf. Eq. (4.2)). The fundamental fact that I demonstrate is that, on a fixed observation point $\boldsymbol{r}_{ob}$, at a given instant near to the moving electron, i.e., to the moving $\hat{\xi}(\Delta\omega)_{m,e}$, the rotational velocity of the ether denoted by $\partial_t\xi_{ob}$, is of the same form as the velocity of a point of a rotating solid. This phenomenon is the electron spin, which, as I show, in the quantum state of an atom, can take only quantized values.

X.2 Electron Spin as a Phenomenon Caused by Ether Elasticity

At the observation point $\boldsymbol{R}_{ob}$ (Cf. Ref. 25), let us consider the electromagnetic field created by a moving electric charge e of velocity $\boldsymbol{V}$. This field derives from the Lienard-Wiechert covariant potential tensor A_μ, for which the expression is

$$A_\mu = -\frac{eV_\mu}{4\pi\varepsilon_0(Rc-\boldsymbol{RV})} \tag{10.1}$$

where V_μ is such that $V_4 = c$ and V_j denote the covariant components of $\boldsymbol{V}$; and where R denotes the radius vector going from the center of the charge to the observation point $\boldsymbol{R}_{ob}$ at the time. If $\boldsymbol{A}$ denotes the vector of covariant spatial components A_j, then the expression for the magnetic field $\boldsymbol{H}$, is

$$\boldsymbol{H} = (\boldsymbol{curlA})/\rho_0. \tag{10.2}$$

We note that the explicit expression for $\boldsymbol{H}$ at $\boldsymbol{R}_{ob}$ is given in (4.15).

Let us consider now the case where $\boldsymbol{R}_{ob}$ is very close to e and denoted $\boldsymbol{r}_{ob}$, i.e., where R, denoted by r, is very small. In this case, in Eq. (4.15)

$$\boldsymbol{r} \wedge [\left(\boldsymbol{r} - \frac{\boldsymbol{V}r}{c}\right) \wedge \dot{\boldsymbol{V}}]/c^2 \text{ is negligible in front of } (\boldsymbol{r} - \boldsymbol{V}r/c)/\beta^2$$

and

$$\boldsymbol{r} \cdot \boldsymbol{V}/\mathrm{c} \text{ is negligible in front of r .}$$

Therefore, at $\boldsymbol{r}_{ob}$, the expression for $\boldsymbol{H}$, denoted $\boldsymbol{H}_{ob}$, is as follows

$$\boldsymbol{H}_{ob} = \frac{e}{4\pi r^3} \boldsymbol{V} \wedge \boldsymbol{r} \,. \tag{10.3}$$

$\frac{e}{4\pi r^3}$ is the density of electrical charge that I denote by ρ_e, i.e., (10.3) can be written

$$\boldsymbol{H}_{ob} = \rho_e \boldsymbol{V} \wedge \boldsymbol{r} \,. \tag{10.4}$$

As shown in Eq. (4.2), the magnetic field $\boldsymbol{H}$ is the speed $\partial_t \boldsymbol{\xi}$ of the rotation of a point of the ether. Therefore, at $\mathbf{r}_{\mathrm{ob}}$ the expression for $\partial_t \boldsymbol{\xi}$, denoted more specifically by $\partial_t \boldsymbol{\xi}_{ob'}$, is

$$\partial_t \boldsymbol{\xi}_{ob'} = \boldsymbol{H}_{ob} \tag{10.5}$$

where $\boldsymbol{H}_{ob}$ is defined in (10.4).

It appears therefore that the velocity $\partial_t \boldsymbol{\xi}_{ob'}$ of an ether point at $\mathbf{r}_{\mathrm{ob}}$ defined by (10.4) and by (10.5) is of the same form as the velocity V_Ω of a point of a rotating solid of rotation vector Ω and of radius r since V_Ω is of the form

$$\boldsymbol{V}_\Omega = \boldsymbol{\Omega} \wedge \boldsymbol{r} \tag{10.6}$$

One sees by considering (10.4) and (10.6) that all happens as if

$$\boldsymbol{\Omega} = \rho_e \boldsymbol{V} \tag{10.7}$$

Remark

$\rho_e \boldsymbol{V}$ has the dimension [1/T] like $[\Omega]$. Indeed, in Ref. 12, I have shown that in the ether elasticity theory $[\rho_e] = [1/L]$.

Let us take $= r_e$, where r_e is the radius of e. In this case, $\boldsymbol{\Omega}$ is the spin of the electron and is denoted $\boldsymbol{\Omega}_e$. In a quantum state, e.g., of a hydrogen atom, we have, for example, Cf. eqs. (9.9), (9.34), and (9.38),

$$V = \frac{e_0 e}{4\pi\varepsilon_0} \frac{1}{n\hbar} \tag{10.8}$$

where q is the positive electric charge of the nucleus, considered to be immobile, around which the electron gravitates. Therefore

$$\boldsymbol{\Omega}_e = \pm \rho_e \frac{e e_0}{4\pi\varepsilon_0} \frac{1}{n\hbar} \tag{10.9}$$

showing the quantum states of electron spin in a hydrogen atom.

CHAPTER XI

THE ELASTIC ETHER THEORY IMPLIES THAT ELECTROMAGNETISM IS THE NEWTONIAN APPROXIMATION OF GENERAL RELATIVITY

XI.1 Introduction

The fundamental Einsteinian tensor $g_{\mu\nu}$, solution of the system of equations (44) of Ref. 19, is usually considered to solely define the gravitational field. However, it is important to note that this system of equations was the result of pure mathematical reasoning and not related, *a priori*, to gravitation. This is the reason why, in my opinion, the title of his famous paper does not contain the word "gravitation". Therefore, we may suppose that these equations and this tensor are more general than just defining the gravitational field, and can define other fields also, such as the electromagnetic field, for example. If this is the case, then Einstein's theory of general relativity would unify the theories of both gravitation and electromagnetism. In this chapter, I show that this is really the case, as folowing. First, the fact that the Coulomb electrostatic potential of force $A_{4,s}$, created by an immobile electrical charge e_0, is of the same form as the Newtonian gravitostatic potential of force $G_{4,s}$, created by an immobile massive particle m_0 implies that *the electrostatic field* $A_{4,s}$ *is a particular gravitational field* $G_{4,s}$ (s for static), in which mm_0k and $-ee_0/(4\pi\varepsilon_0)$ are equivalent.

The question remains as to whether the electromagnetic potential tensor A_μ, created by the moving_e_0, is of the same form as a particular gravitational tensor G_μ, created by the moving m_0.

The answer is yes and here are the qualitative lines of this affirmation demonstrated here below and also in Ref. 26.

In Sec. IV.6 and in Sec. VII, I have shown that the moving m at the velocity $\boldsymbol{V}$ where m can be zero, in this case it is a photon, is associated a rotating ether trajectory segment, $\boldsymbol{R}(\phi_P, \phi)$, of limited length in which the points ether rotate and the rotational axes follow the trajectory. The interesting point is that these rotating points induce rotations in their neighbors, which go on to induce rotations in their neighbors, and so on. These induced rotations are, in fact, the field created by this rotating ether trajectory segment. For example, the field created by an electric charge e in its motion is the magnetic field $\boldsymbol{H}$ in which the rotational axes of the ether points go in the same direction as the trajectory of the rotating ether trajectory segment. The relevant ensemble of such rotations creates a field $\boldsymbol{H}$ around the trajectory of this charge. This field is circular when the electron trajectory is a straight line and simply composed by closed trajectories in the non-straight electron trajectories.

Outside of this $\boldsymbol{R}(\phi_P, \phi)$, the ether points rotate and these axes of rotation are parallel to the motion, forming the magnetic field $\boldsymbol{H}$ around the trajectory of $\boldsymbol{R}(\phi_P, \phi)$. The perturbation caused by this motion is propagated in the free ether with velocity c, and is a propagated gravitational field and a magnetic field, since it is created by m and the electric charge e. The Newtonian approximation of this propagated field is a tensor G_μ, obtained by using the Lorenz transform with $\boldsymbol{V}$ as a constant (so as not to involve other fields) and $G_{\mu,s}$. It appears that G_μ is a_Lienard-Wiechert potential_tensor, like the electromagnetic potential tensor A_μ created by the moving e. Furthermore, I show that changes of the ether caused by e is of the same form as that caused by m, and that the interaction of such deformations are the interaction forces.

XI.2 Notations and Reminders

The Lagrange-Einstein function L_G of a particle of mass m and electric charge e, subjected to a gravitational field $g_{\mu\nu}$ and to an electromagnetic potential tensor A_μ, is defined in (5.1), from which we can deduce the motion equation (Cf. Sec. V.2). When this particle is subjected only to A_μ, then L_G becomes $L_{G,EM}$, such that

$$L_{G,EM} = -mc\sqrt{c^2 - V^2} + qA_\mu \dot{x}^\mu / c \,. \qquad (11.1)$$

If the electromagnetic field is due to the immobile electric charge e_0 at **O**, then $L_{G,EM}$ becomes $L_{G,EMS}$, such that

$$L_{G,EMS} = -mc\sqrt{c^2 - V^2} - \frac{ee_0}{4\pi\varepsilon_0 r} \,. \qquad (11.2)$$

If a particle of mass m is subjected only to a gravitational field $g_{\mu\nu}$, then L_G is denoted $L_{G,G}$, for which the expression is

$$L_{G,G} = -mc\sqrt{g_{\mu\nu}\dot{x}^{\mu}\dot{x}^{\nu}}\,. \tag{11.3}$$

XI.3 Newtonian Approximation of the Lagrange-Einstein Function of a Massive Particle in a Gravitation Field

The explicit form of $g_{\mu\nu}\dot{x}^{\mu}\dot{x}^{\nu}$ is

$$g_{\mu\nu}\dot{x}^{\mu}\dot{x}^{\nu} = g_{\mu\mu}\dot{x}^{\mu}\dot{x}^{\mu} + 2g_{4j}\dot{x}^{4}\dot{x}^{j} + 2\Delta_{i\neq j} \tag{11.4}$$

where $\Delta_{i\neq j}$ is defined by

$$\Delta_{i\neq j} \equiv g_{12}\dot{x}^{1}\dot{x}^{2} + g_{13}\dot{x}^{1}\dot{x}^{3} + g_{23}\dot{x}^{2}\dot{x}^{3} \tag{11.5}$$

and $g_{\mu\mu}$ as following

$$g_{\mu\mu} = g_{0,\mu\mu} + \delta g_{\mu\mu} \tag{11.6}$$

for which $g_{0,\mu\mu}$ denotes the **free** value of $g_{\mu\nu}$. With these notations, (11.4) becomes

$$g_{\mu\nu}\dot{x}^{\mu}\dot{x}^{\nu} = c^2 - V^2 + \delta g_{\mu\mu}\dot{x}^{\mu}\dot{x}^{\mu} + 2g_{4j}\dot{x}^{4}\dot{x}^{j} + 2\Delta_{i\neq j} \tag{11.7}$$

and (11.3) can be written in the form

$$L_{G,G} = -mc\sqrt{c^2 - V^2} - \frac{mc}{2\sqrt{c^2-V^2}}\left(\delta g_{\mu\mu}\dot{x}^{\mu}\dot{x}^{\mu} + 2g_{4j}\dot{x}^{4}\dot{x}^{j} + 2\Delta_{i\neq j}\right) + \cdots\,. \tag{11.8}$$

If the incident gravitational field is created by a particle of immobile mass m_0 located at O ($m_0 >> m$), i.e., when this incident gravitational field is of a Schwarzschild form, then $L_{G,G}$, denoted $L_{G,GS}$, becomes

$$L_{G,GS} = -mc\sqrt{c^2\gamma^2 - V^2\hat{\gamma}^2}\,. \tag{11.9}$$

Now (11.9) can be written as

$$L_{G,GS} = -mc\sqrt{c^2 - V^2} + \frac{mc\alpha}{2r\sqrt{c^2-V^2}}\left(c^2 + \frac{V^2\cos^2 a}{\gamma^2}\right) + \cdots\cdots \tag{11.10}$$

Considering (11.8), we can see that the Newtonian approximation of $L_{G,G}$, denoted $L_{G,G,NA}$, is

$$L_{G,G,NA} = -mc\sqrt{c^2 - V^2} + mG_\mu \dot{x}^\mu / c \qquad (11.11)$$

where G_μ is the tensor defined by

$$G_4 = -c^2 \delta g_{44}/2 \,, \qquad G_j = -c^2 \delta g_{4j} \quad . \qquad (11.12)$$

It appears that eqs. (11.1) and (11.11) are of the same form and therefore we can suppose that the tensor qA_μ plays the same role as the tensor mG_μ. This *supposition* is reinforced by considering the static Newtonian approximation of Eq. (11.9), or by the fact that, in the static case, $\delta g_{44} = -\alpha/r$, i.e., the expression for $G_{4,S}$, (S for stationary), is

$$G_{4,S} = m_0 k/r \,. \qquad (11.13)$$

Therefore, in the static case, $L_{G,G,NA}$ denoted $L_{G,G,NAS}$, becomes

$$L_{G,G,NA} = -mc\sqrt{c^2 - V^2} + mm_0 k/r \,. \qquad (11.14)$$

We can see that (11.2) is of the same form as (11.14), and that $-\mathrm{qq}_0/4\pi\varepsilon_0$ and mm_0k are equivalent. This shows that the mass m can theoretically be negative (this is a subject that needs a special frame of discussion). As such, the electrostatic field created by an immobile electric charge, i.e., a Coulomb field, differs from the Newtonian approximation of the gravitational field created by an immobile point mass, i.e., from the Newtonian approximation of the Schwarzschild field, by only a constant coefficient of proportionality. Therefore, it appears that:

The electrostatic field is a particular case of stationary gravitation, $-\frac{ee_0}{4\pi\varepsilon_0 r}$ *and* mm_0k *being equivalent, i.e., playing the same role.*

XI.4 The Rotating Ether Trajectory Segment Associated with the Mobile Massive Particle: Its Behavior When Immobile

Let us consider a moving point, mass m_0, of velocity $\boldsymbol{V}_{m_0}$, considered constant for the sake of simplicity and directed along the x axis. In Sec. VII.2, I showed, in particular, that a "single particle wave" $\hat{\boldsymbol{\xi}}(\Delta\omega)$ is associated with this point mass for which the expression is

$$\hat{\boldsymbol{\xi}}(\Delta\omega) = \boldsymbol{\xi}_0 \exp[i\omega(-t + \frac{x}{V_P})]SINC[\frac{\Delta\omega}{2}(-t + \frac{x}{V_{m_0}})] \qquad (11.15)$$

which is the packet of waves defined in (7.7). In Eq. (11.15), the function $\mathrm{SINC}(y)$ is defined by $\mathrm{SINC}(y) \equiv (\sin y)/y$, $\boldsymbol{\xi}_0$ denotes a vector that only depends on the spatial coordinates, V_P, the phase velocity for which the expression is

$$V_P = c/\sqrt{1 - (m_0c^2/E_T)^2} \qquad (11.16)$$

ω is the angular frequency defined by $\omega \equiv E_T/\hbar$, and E_T is the particle's total energy. Furthermore, the expression for V_{m_0} is defined in Eq. (5.59). Equation (11.15) shows that:

> $\hat{\boldsymbol{\xi}}(\Delta\omega)$ *is a packet of waves, i.e., forming* $\boldsymbol{R}(\phi_P, \phi)$, *which moves with the velocity* V_{m_0}, *in which the ether vibrates at the frequency* $\omega/(2\pi)$ *and the larger* V_{m_0} *is, the smaller* $\boldsymbol{R}(\phi_P, \phi)$ *is.*

In order to be more explicit, $\hat{\boldsymbol{\xi}}(\Delta\omega)$ is also denoted $\hat{\boldsymbol{\xi}}(\Delta\omega, m_0, \boldsymbol{V}_{m_0})$.

For an immobile free point of mass m_0, then $E_T = m_0c^2$, (Cf. (11.16)) V_P is infinite, and $\hat{\boldsymbol{\xi}}(\Delta\omega, m_0, 0)$ associated with this immobile particle is

$$\hat{\boldsymbol{\xi}}(\Delta\omega, m_0, 0) \equiv \boldsymbol{\xi}_0 \exp\left(-i\frac{2\pi}{h}mc^2t\right) SINC\left[\frac{\Delta\omega}{2}\left(-t + \frac{x}{0}\right)\right]. \qquad (11.17)$$

Since $SINC(\infty) = 0$, it follows that $\hat{\boldsymbol{\xi}}(\Delta\omega, m_0, 0) = \boldsymbol{0}$. The question remains as to what has changed this $\boldsymbol{R}(\phi_P, \phi)$. The answer is:

The immobile point mass m_0 *is gravitational, it creates a Schwarzschild field, which, in the Newtonian approximation, is of the same form as an electrostatic field.*

XI.5 The Field Created by a Moving Electric Charge as Similar to the Newtonian Approximation of the Field Created by a Moving Massive Particle

$\hat{\boldsymbol{\xi}}(\Delta\omega, m_0, \boldsymbol{V}_{m_0})$, defined in (11.15), is a moving perturbation of the ether associated with a particle of mass m_0 and velocity $\boldsymbol{V}_{m_0}$, as defined in Eq. (5.59). This perturbation is $\boldsymbol{R}(\phi_P, \phi)$ of limited volume. It can be small and moves also with the velocity $\boldsymbol{V}_{m_0}$. Outside of this globule, *the ether is perturbed only by the motion of this* $\boldsymbol{R}(\phi_P, \phi)$. This ether perturbation, denoted $P[\hat{\boldsymbol{\xi}}(\Delta\omega, m_0, \boldsymbol{V}_{m_0})]$ and due only to the motion of $\hat{\boldsymbol{\xi}}(\Delta\omega, m_0, \boldsymbol{V}_{m_0})$, is propagated in the free ether, i.e., at velocity c, which is much greater than $\boldsymbol{V}_{m_0}$. This is to say that, after its emission, $P[\hat{\boldsymbol{\xi}}(\Delta\omega, m_0, \boldsymbol{V}_{m_0})]$ is no longer influenced by $\hat{\boldsymbol{\xi}}(\Delta\omega, m_0, \boldsymbol{V}_{m_0})$. I now determine $P[\hat{\boldsymbol{\xi}}(\Delta\omega, m_0, \boldsymbol{V}_{m_0})]$ at a fixed observation point $\boldsymbol{R}_{ob}$. To this end, I use the fact that the Newtonian approximation (NA) of $[\hat{\boldsymbol{\xi}}(\Delta\omega, m_0, 0)]$, i.e., that of the ether perturbation due to the immobile m_0, is no other than $G_{4,S}$, as defined in (11.13). Therefore, the NA of $P[\hat{\boldsymbol{\xi}}(\Delta\omega, m_0, \boldsymbol{V}_{m_0})]$ is the tensor G_μ obtained by applying the Lorentz transform, with $\boldsymbol{V}_{m_0}$ as a constant and $G_{4,S}$. $\boldsymbol{V}_{m_0}$ is taken to be constant so as not to involve other fields. Let us now consider this Lorentz transform. Using the notation

$$\beta_0 = 1/\sqrt{1 - \left(V_{m_0}/c\right)^2} \qquad (11.18)$$

this Lorentz transform, defined by the equations (38.1), (38.2), and (38.4) (Ref. 15), yields

$$G_\mu = \frac{1}{c}\beta_0 G'_{4,S} V_{m_{0,\mu}} = \frac{m_0 k \beta_0 \boldsymbol{V}_{m_{0,\mu}}}{cR'} \qquad (11.19)$$

where $V_{m_{0,j}}$ denotes the covariant component "j" of $\boldsymbol{V}_{m_0}$, $V_{m_{0,4}} = c$ and R'/β_0 is the quantity defined by

$$R'/\beta_0 = \sqrt{(x - V_{m_0}t)^2 + (y^2 + z^2)/{\beta_0}^2}\,. \qquad (11.20)$$

Considering that $t = R/c$, where $R = \sqrt{x^2 + y^2 + z^2}$, the right-hand member of (11.20) becomes

$$R - xV_{m_0}/c$$

and since $\boldsymbol{V}_{m_0}$ is parallel to the x axis, i.e.,

$$xV_{m_0} = \boldsymbol{R} \cdot \boldsymbol{V}_{m_0}$$

then, (11.20) becomes

$$R'/\beta_0 = R - \boldsymbol{R} \cdot \boldsymbol{V}_{m_0}/c \,. \tag{11.21}$$

Inserting (11.21) into (11.19), we obtain

$$G_\mu = m_0 k V_{m_{0,\mu}}/(Rc - \boldsymbol{R} \cdot \boldsymbol{V}_{m_0}) \,. \tag{11.22}$$

Equation (11.22) shows that G_μ is a Lienard-Wiechert potential tensor, i.e., it is the gravitational potential field seen at an observation point R_{ob} due to a particle of mass m_0 that moves with velocity $\boldsymbol{V}_{m_0}$. Here R is the distance between the position of m_0 at time t' where the signal was emitted, reaching point R_{ob} at time t such that $(t - t')c = R$.

The electromagnetic field created by a moving electric charge q_0, of velocity $\boldsymbol{V}_{q_0}$, that reaches point R_{ob} is the electromagnetic Lienard-Wiechert potential A_μ, defined by

$$A_\mu = -q_0 V_{q_{0,\mu}}/[4\pi\varepsilon_0(Rc - \boldsymbol{R} \cdot \boldsymbol{V}_{q_0})] \tag{11.23}$$

where the $V_{q_{0,j}}$ is the covariant components of $\boldsymbol{V}_{q_0}$ and $V_{q_{0,4}} = c$. When $\boldsymbol{V}_{q_0} = 0$, then only A_4 is not null. It is the denoted $A_{4,s}$, for which the expression is

$$A_{4,s} = -q_0/(4\pi\varepsilon_0\, R) \tag{11.24}$$

i.e., it is the Coulomb potential, of the same form as $G_{4,s}$ defined in (11.13). A_μ of (11.23) is a solution of the equation

$$\nabla^2 A_\mu - \partial_{44} A_\mu = -q_0 V_{q_{0,\mu}}/(c\varepsilon_0) \,. \tag{11.25}$$

and $A_{4,s}$ of the Laplace equation $\nabla^2 A_{4,s} = -q_0/\varepsilon_0$. Therefore, G_μ is a solution of the equation

$$\nabla^2 G_\mu - \partial_{44} G_\mu = 4\pi m_0 k V_{m_{0,\mu}}/c \tag{11.26}$$

and $G_{4,s}$ of the Laplace equation $\nabla^2 G_{4,s} = 4\pi m_0 k$. Considering (11.11), (11.12), and (11.22), the expression for the NA of the Lagrange-Einstein function $L_{G,G,NA}$ is

$$L_{G,G,NA} = -mc\sqrt{c^2 - V^2} + \frac{mm_0\, k}{c(Rc - \boldsymbol{R}\cdot\boldsymbol{V}_{m_0})}\dot{x}^\mu V_{m_{0,\mu}} \qquad (11.27)$$

while considering (11.1) and (11.23), the expression for $\mathrm{L}_{\mathrm{G,EM}}$ is

$$L_{G,EM} = -mc\sqrt{c^2 - V^2} - \frac{ee_0}{4\pi\varepsilon_0 c(Rc - \boldsymbol{R}\cdot\boldsymbol{V}_{q_0})}\dot{x}^\mu V_{q_{0,\mu}} \qquad (11.28)$$

where $\dot{x}^\mu$ is related to the particle subjected to the field. For $\dot{x}^j \ll c$, (11.27) is very close to (11.14), and (11.28) to (11.2).

We can see the similitude of an ether perturbation due to the electric charge q_0 *with the Newtonian approximation of the one due to a massive particle* m_0, *that is to say, the similitude of the electromagnetic Lienard-Wiechert potential tensor* (11.23) *with the gravitational Lienard-Wiechert potential tensor* (9.22), *and the similitude of the electromagnetic Lagrange-Einstein function* $L_{G,EM}$ (11.28) *with the Newtonian approximation of the gravitational Lagrange-Einstein function* $L_{G,G,NA}$ (11.27).

XI.6 A Particular Solution for Stokes' Formula

Stokes' formula is

$$\int_C \xi d\ell = \iint_{S(C)} \boldsymbol{curl\xi}\cdot\boldsymbol{n}dS \qquad (11.29)$$

where $\boldsymbol{\xi}$ is a spatial vector, C is a closed curve, and S(C) a surface bounded by C. Let us take

$$\boldsymbol{curl\xi} = b_1\,\boldsymbol{r}/r^3 \qquad (11.30)$$

where b_1 is a constant, and consider the particular case where C is a small circle of radius R_C perpendicular to $\boldsymbol{r}$, such that $r >> R_C$. In terms of symmetry, $\boldsymbol{\xi}$ is parallel to $d\ell$ and, therefore, (11.29) becomes

$$\int_C \boldsymbol{\xi} d\ell = 2\pi R_C \boldsymbol{\xi} \,. \qquad (11.31)$$

Furthermore, considering again that $r >> R_C$, using a simple calculation, we obtain

$$\iint_{S(C)} \frac{\boldsymbol{r}}{r^3} \cdot \boldsymbol{n} dS \cong \pi \left(\frac{R_C}{r}\right)^2 . \qquad (11.32)$$

From (11.31) and (11.32), we can deduce that, on C of radius R_C, we have

$$\xi = b_1 R_C/(2r^2) \qquad (11.33)$$

for which the vectorial form is

$$\boldsymbol{\xi} = \frac{b_1}{2r^3} \boldsymbol{r} \wedge \boldsymbol{R}_C \,. \qquad (11.34)$$

Let us consider the case where, in Cartesian coordinates, $\boldsymbol{r}$ follows the z axis and $\boldsymbol{R}_C$ is the vector defined by

$$\boldsymbol{R}_C \equiv R_C[cos\Phi,\ sin\Phi,\ 0] \,. \qquad (11.35)$$

In this case

$$\boldsymbol{\xi} = \frac{b_1}{2r^2} R_C[cos\Phi,\ sin\Phi,\ 0] \qquad (11.36)$$

and the angle Φ, described by a point of the circle C, is $\Phi = \xi/R_C$, i.e., in vector form

$$\boldsymbol{\Phi} = \frac{\boldsymbol{R}_C}{R_C} \wedge \frac{\boldsymbol{\xi}}{R_C} = \frac{b_1}{2r^3} \boldsymbol{r} \,. \qquad (11.37)$$

XI.7 The Electromagnetic Force and Gravitational Force as the Interaction of the Ether Changes Due to Electric Charges, Resp. to Massive Neutral Particles

The electrostatic field $\boldsymbol{E}_{q_0}$ due to e_0 at **O** is related to the rotations $\boldsymbol{\xi}_{e_0}$ of the points of the ether by

$$\boldsymbol{curl\xi}_{e_0} = \varepsilon_0 \boldsymbol{E}_{e_0} \qquad (11.38)$$

(Cf. Eq. (5) of Ref. 13).

Considering (11.2) and (11.38), we can obtain the expressions for $\boldsymbol{curl\xi}_{e_0}$, ξ_{e_0} and for Φ, denoted now Φ_{e_0}, by replacing b_1 by $e_0/(4\pi)$, $\boldsymbol{R}_C$ also being defined by (11.35) but with Φ_{e_0} instead of Φ. Thus, from (11.37), we have

$$\boldsymbol{E}_{e_0} = 2\eta_0 \boldsymbol{\Phi}_{e_0} = e_0 \boldsymbol{r}/(4\pi r^3) \,. \qquad (11.39)$$

where $\eta_0 \equiv 1/\varepsilon_0$. Let $\boldsymbol{F}_{e_0,e}$ be the interactive force between e_0 and another electric charge e. As demonstrated in Ref. 13, the expression for the potential $U_{e_0,e}$, such that $\boldsymbol{F}_{e_0,e} = -\boldsymbol{\nabla}(U_{e_0,e})$, is given by Eq. (26) (ibid) and (11.39), namely by

$$U_{e_0,e} = \frac{e_0 e}{4\pi r} = \varepsilon_0 \iiint_{B(0,r)} \mathbf{E}_{q_0} \mathbf{E}_q dV_{ol} = \frac{4}{\varepsilon_0} \iiint_{B(0,r)} \mathbf{\Phi}_{q_0} \mathbf{\Phi}_q dV_{ol} \quad (11.40)$$

where B(0,r) denotes the sphere of radius r centered at **O** and dV_{ol} is the element of volume.

The NA of the gravitostatic field $\boldsymbol{E}_{m_0,NA,S}$ due to an immobile neutral massive particle of mass m_0 is

$$\boldsymbol{E}_{m_0,NA,S} = -m_0 k \boldsymbol{r}/r^3 \,. \qquad (11.41)$$

This is due to the fact that, in the NA, the force $\boldsymbol{F}_{m_0,NA}$ acting between m_0 and another particle of mass m is

$$\boldsymbol{F}_{m_0,m,NA} = -\frac{m m_0 k \boldsymbol{r}}{r^3} = m\boldsymbol{E}_{m_0,NA,S} \,. \qquad (11.42)$$

This force is derived from the potential of force $U_{m_0,m}$, defined by

$$U_{m_0,m} = -\frac{m m_0 k}{r} \qquad (11.43)$$

since

$$\boldsymbol{F}_{m_0,m,NA} = -\boldsymbol{grad} U_{m_0,m} \,.$$

Using the same calculation as in Sec. 4.2 of Ref. 13, and comparing the result with (11.43), we get

$$U_{m_0,m} = -\frac{1}{k}\iiint_{B(0,r)} \boldsymbol{E}_{m_0}\boldsymbol{E}_m dV_{ol} \tag{11.44}$$

Now

$$\boldsymbol{curl\xi_{m_0}} = m_0\boldsymbol{r}/r^3 \tag{11.45}$$

that is to say

$$2\boldsymbol{\Phi}_{m_0} = \boldsymbol{curl\xi_{m_0}} \tag{11.46}$$

and therefore

$$\boldsymbol{E}_{m_0,NA,S} = -2k\boldsymbol{\Phi}_{m_0} \,. \tag{11.47}$$

It follows, by comparison with (11.40), that

$$U_{m_0,m} = -mm_0\frac{1}{r} = -4k\iiint_{B(0,r)} \boldsymbol{\Phi}_{m_0}\cdot\boldsymbol{\Phi}_m dV_{ol} \,. \tag{11.48}$$

This shows us that:

> *the electrostatic field $\mathbf{E}_{q_0}$ and the gravitostatic field $\mathbf{E}_{m_0,NA,S}$ are densities of couples of forces applied to the ether, and the inductions $\varepsilon_0\mathbf{E}_{q_0}$ and $-\boldsymbol{E}_{m_0,NA,S}/k$ are twice the angle of rotation of the ether around the radius vector $\boldsymbol{r}$ caused by the electric charge q_0, Resp. by a massive particle m_0, at **O**.*

CHAPTER XII

INTERPRETATION OF THE "CHEMICAL" VS. "NUCLEAR" REACTION, OF THE ATOM'S STABILITY OR INSTABILITY, AND OF THE OXIDIZED ATOM AND THE NOBLE ATOM THROUGH THE THEORY OF ETHER ELASTICITY

XII.1 Interpretation of the "Chemical" Vs. "Nuclear" Reaction and of the Atom's Stability or Instability through the Theory of Ether Elasticity

A chemical reaction between several atoms results in the same number of atoms, but now connected forming a unique molecule. For example, one atom of oxygen O and two atoms of hydrogen H form a molecule of water $(2H)O$. Thus, a chemical reaction is an electronic one, since, after the reaction, the number of atoms remains the same, that is, all the nuclei involved in this reaction remain separated.

In contrast to a chemical reaction, a nuclear reaction also occurs between several atoms and results in a unique atom such that all the nuclei of these atoms are fused into a single nucleus. For example, from a nuclear reaction involving two H atoms, the result is one atom of helium H_e^2 the nucleus of which being constituted by the two H nuclei and neutrons.

In a nuclear reaction (Cf. Sec. VII), protons attract each other to form a new nucleus. More precisely, as has been shown to be due to the Schwarzschild attraction, in so-called "strong gravitation", two protons are attracted to each other once they are sufficiently close to one another, but repulse one another when their distance is greater than the threshold. This explains why two atoms of hydrogen can become a helium atom, i.e., a stable atom, while heavy atoms, i.e., those having a large number of protons, cannot. This is because some distances between the protons are greater than the distance at which their attraction supersedes their repulsion. Therefore, a nuclear attractive reaction is none other than strong gravitation, as show in Sec. VII above.

XII.2 The Oxidized Atom and the Noble Atom Ensuing from Ether Elasticity

An electron is constituted by a sum of monochromatic waves (Cf. Eq. (7.7)), but when this electron is confined in an atom, it is only one of these monochromatic waves. This wave is called a "resonant electron wave". The frequency of its wavelengths around the nucleus is an integer, while the other monochromatic waves, the frequencies of which around the nucleus are not integers, auto-destruct through interference.

An atom is said to be **noble** when each of its resonant waves is coupled with another electron wave propagating along the same trajectory, but in the opposite sense. An atom is said to be **oxidized** when it has at least one remaining resonant electron wave that is not coupled with another electron wave (Cf. [6]).

In a nuclear reaction, the resulting atom can be noble or oxidized, the properties of which can be defined in the frame of the ether elasticity theory.

In a noble reaction, each such pair creates a standing wave of energy that is much smaller than the sum of their individual energies. The difference in energy is emitted in the form of electromagnetic energy. For example, in the creation of the atom H_e^2, the two electron waves H that create this H_e^2 propagate in opposite directions around the new nucleus formed by the two H nuclei. As such, we have to specifically denote each of these H atoms, for example, by H_+ and H_-. The creation of the following noble atoms occurs in the same way: N_e^{10}, A_r^{18}, and K_r^{36}.

In contrast, in an oxidized reaction, some electron waves of the atom are not coupled with another electron wave, even when the number of electrons in the atom is a "pair number" like, for example, the carbon atom, which has 6 electron waves, but not three stationary waves. It is clear that an atom possessing an unpairable number of electrons cannot be noble.

The question remains as to why, for example, the carbon atom is not noble, that is, why does it not possess three stationary waves? An analogous question holds regarding other atoms that also possess a pair number of electrons, like beryllium and oxygen, etc., but that do not have only stationary waves. The answer lies in quantum analysis. It appears that *if all the atoms possessing a pair number of electrons were noble, then it would reduce by half the possibility of chemical reactions, and therefore the nature of our world would be different.*

XII.3 Mathematical Expressions for the Noble Atom and the Oxidized Atom

We now mathematically consider the noble atom. As stated above, a noble atom is constituted by a pair number of electrons, the waves of each pair propagating on the same trajectory around the nucleus, but in opposite directions, i.e., Cf. eq. (9.45) of the form

$$\boldsymbol{P}(0, \boldsymbol{V}_{Pc}) = (1+N)\xi_0 sin(-\omega t + \theta n_0) \quad (12.1)$$

$$\boldsymbol{P}(0, -\boldsymbol{V}_{Pc}) = (1+N)\xi_0 sin(-\omega t - \theta n_0)\ . \quad (12.2)$$

I demonstrate now that two such waves interfere and create a stationary wave. Indeed, a pair of such electron waves propagating in opposite senses around the nucleus at the same frequency form a stationary wave the total kinetic energy of which is much smaller than the sum of the non-coupled kinetic energy. Indeed, the two electron waves defined in (12.1) and (12.2) build a stationary wave, since their sum $\Delta\xi$ defined by

$$\Delta\xi \equiv \boldsymbol{P}(0, \boldsymbol{V}_{Pc}) - \boldsymbol{P}(0, -\boldsymbol{V}_{Pc}) \quad (12.3)$$

is

$$\Delta\xi = 2(1+N)\xi_0 cos(\omega t) sin(\omega \ell / V_{Pc}) \quad (12.4)$$

which, on the circle of radius r, i.e., where $\ell \equiv r\theta$, becomes

$$\Delta\xi = 2(1+N)\xi_0 cos(\omega t) sin(\omega r\theta / V_{Pc})\ . \quad (12.5)$$

This is zero at

$$\theta = KV_{Pc}/2r\nu \quad (12.6)$$

where K is an integer. Furthermore, with λ denoting the wavelength defined by $V_{Pc} = \nu\lambda$, the angle $\Delta\theta$ between two zeros is

$$\Delta\theta = \lambda/(2r)\ . \quad (12.7)$$

A consequence of this expression is that the angular distance $\Delta\theta$ between two zeros of a stationary wave around the nucleus can be very small. In this case, on each circuit the particle emits a very small amount energy that is compensated for by the fact that the wave has a significant length i.e., N is very large and theoretically infinite, and therefore it superposes itself an additional N times. This explains the atom's stability, i.e., why in noble cases the electrons gravitating around the nuclei do not fall onto them and the arrangement is much more stable than in oxidized cases.

We may recall that the oxidized case is the one for which the expression for the electron defined, in general, by Eq. (7.7) becomes simply eqs. (12.1) or (12.2), but not their sum, that is, a single frequency phase wave. Thus, when all the electron waves only form such stationary waves in an atom, it is said to be noble. It is evident that an atom of unpaired numbers of electrons, like, for example, the nitrogen atom, cannot be noble, since not all of its layers can be saturated. An atom possessing a pair number of electrons is described as being oxidized when there are wave electrons that do not interfere with one another.

This justification follows directly from the theory of the ether, as otherwise we would not be able to understand why, in certain cases, two electrons have an energy smaller than the sum of their independent energies.

CHAPTER XIII

CONCLUSIONS

In this book, I have shown that the elasticity theory, in which the medium is shown to be an elastic medium governed **by the "appropriated Navier-Stokes-Durand equation"** unifies the theories of physics. By the word "appropriated", I mean that the Navier-Stokes-Durand equation, being a general equation of elasticity, is restricted to the case that contains the fundamental component that unifies the different theoretical domains of physics. As such, this appropriated Navier-Stokes-Durand equation follows from the general theory of elasticity with the difference that I consider that the points of this medium do not move linearly, one relative to the other, but can only rotate on themselves. These rotations are caused by point couples of forces denoted C, applied to this medium, and, also, where this medium is not subjected to any linear forces. These point rotations are then transmitted, point-to-point, by the fact that a rotating point in the medium induces rotations in its neighboring points, resulting in the propagation of fields and particles. The fact that these elastic medium points do not move linearly is due to the fact that, if there were such linear motions, then space would be linearly deformed, for example, near what is called the Schwarzschild horizon, the spatial contraction would be infinite in this case and there would be no order in space. It follows that **the only changes that points in the medium can undergo are rotations of linearly immobile medium points. The medium is called the ether.**

In free space, there is no vacuum, but there is a specific medium called the ether. The concept of the ether, envisaged in particular by Maxwell and Einstein (Cf. refs. 7-9), and developed further by the author (Cf. refs. 10-11), has been presented here in a detailed manner. It appears, as I show, that the ether is an elastic medium in which its points can **only rotate elastically on themselves**. By the term "elastically", I mean that when the cause of this rotation no longer exists, then the rotation disappears and the ether point returns to its free state. It appears that, contrary to some opinions, this concept of the ether allows us to define the fundamental notions of dimensions, time, and velocity, and to unify the fundamental notions of physics. Indeed, the ether is a specific elastic isotropic medium of constant properties in which one can define point positions and the "distance" between any two fixed points. This important fact, namely, that one can define the point positions and their distances from one another, has been analyzed, since the medium of the ether has to be defined with specific characteristics.

Light has been shown to be a kind of perturbation created in a given ether domain and propagated in the ether out of this perturbing region. If the ether is not perturbed by another cause, then light always propagates in the same manner. It is the fact that the ether is an elastic isotropic medium of constant properties that allows us to define the fact that light propagates in non-perturbed ether always in the same manner, and, as has been shown, *specific rotational ether segments* can be formed in the ether.

I have proven that there are two different kind of velocity—the phase velocity and the group velocity—which are related to each other. Examples of these two kinds of velocity have been presented, along with their influence on their environments. I have shown that that what is called a particle is, in fact, a rotating ether trajectory segment, being formed by a sum of waves of close phase velocities.

I recalled the demonstration, which is a continuation of Einstein's special relativity theory, concerning the results of the Morley-Michelson experiment. In this demonstration, I have shown how this experiment confirms the fact that there is *no change* in ether density, but rather there is a change in the distance *travelled by the photon* in the moving reference, since, in this experiment, two photons of different trajectories meet.

It is known that, for the *Newton-Maxwell case*, electrical repulsion between two massive electrically charged particles of the same sign is greater than their Newtonian attraction.

However, as I have demonstrated above, **by taking into account general relativity and the ether theory, their mutual gravitational attraction *surpasses* their electrical repulsion when they are *sufficiently close one to another*.**

The univocal motion of a classical massive particle occurs when there is a *full superposition of waves*, forming a single particle wave $\boldsymbol{R}(\phi_P, \phi)$. However, such a full superposition *cannot exist in a bound state* at the atomic level, such as, for example, an electron in an atom. In such a bound state, the waves that constitute this *superposition interfere with themselves*, and only the resonant ones remain undestroyed by this interference and may even be amplified. These resonant waves, i.e., "quantum states, can be directly determined even in the presence of gravitation.

A specific sum of waves $\xi_{m,e}$ forms a "rotating ether trajectory segment" $\boldsymbol{R}(\phi_P, \phi)$ that moves like $Par(m, e)$ and contains all its parameters. In its motion, this $Par(m, e)$, i.e., $\boldsymbol{R}(\phi_P, \phi)$, creates a Lienard-Wiechert covariant potential tensor from which we can deduce the electromagnetic field, and, in particular, the magnetic field $\boldsymbol{H}$.

In the ether elasticity theory, *this* $\boldsymbol{H}$ *is the rotation velocity* $\partial_t \boldsymbol{\xi}$ *of ether points*. The fundamental fact that I have demonstrated, is that, on a fixed observation point $\boldsymbol{r}_{ob}$ at a given instant near to the moving electron, i.e., to the moving $\boldsymbol{R}(\phi_P, \phi)$, the rotation velocity of the ether, denoted $\partial_t \boldsymbol{\xi}_{ob}$, is of the same form as the velocity of a point of a rotating solid. This phenomenon is electron spin, which, as I have shown, in the quantum state of an atom, can take only quantized values.

In a chemical reaction between several atoms, the result has the same number of atoms, but these are now connected forming a unique molecule.

In contrast to a chemical reaction, a nuclear reaction also occurs between several atoms and results in a unique atom such that *all the nuclei of these atoms have been fused into a single nucleus.* In a nuclear reaction (Cf. Sec. VIII), protons are attracted to each other and form a new nucleus. More precisely, as has been shown to be due to Schwarzschild attraction, in so-called strong gravitation, two protons are attracted to each other when they are sufficiently close together, but repulse one another when their distance is greater than the threshold. These facts explain why two atoms of hydrogen can become a helium atom, i.e., a stable atom, while heavy atoms, i.e., those with a large number of protons, cannot be stable because some of the distances between the protons are greater than the distance at which their attraction supersedes their repulsion.

REFERENCES

1. J. C. Maxwell, *A Treatise on Electricity and Magnetism*, (Dover, New-York, 1954)

2. A. Einstein, *Zur Elektrodynamik bewegter Körper*, Annalen der Physik, vol. 17, pp. 891-921 (1905). A. Einstein, Ether and the theory of relativity: The collected papers of Albert Einstein (Springer, Berlin, 1920), address delivered on May 5, 1920, at the university of Leyden

4. E. Whittaker, *A History of the Theories of Aether and Electricity*, (Thomas Nelson and Sons Ltd, London Edinburgh Paris Melbourne, Toronto and New York, 1961)

5. A. Sommerfeld, *Lectures on Theoretical Physics*, Vol. II "*Mechanics of Deformable Bodies*", (Academic Press Inc., New York, 1950)

6. A. Einstein, *Sci. Am. Special Issue*, 41-45 (1991) (reprinted from the April 1950 issue)

7. A. Einstein, "On the electrodynamics of moving bodies", published in Ref. 8

8. H. A. Lorentz, H. Weyl, H. Minkowski, A. Einstein, "The principle of relativity", (Dover, New York, 1952)

9. A. Einstein, "On the generalized theory of gravitation", *Sci. Am. Spec. Issue*, 40-45 (1991) (reprinted from the April 1950 issue)

10. D. Zareski, "On the elasto-undulatory interpretation of fields and particles". Physics Essays Publication, 25, 2 (2012)

11. D. Zareski, "The elastic interpretation of electrodynamics", Foundation of Physics, Vol. 14, No. 5. October 2001

12. E. Durand, *Electrostatique et Magnétostatique*. Cf. pp. 229-239, & 745-748, (Masson, Paris, 1953)

13. D. Zareski, "Fields and wave-particle reciprocity as changes in an elastic medium: the ether". Phys. Essays 26, 2 (2013)

14. D. Zareski, "The elastic nature of the mass and of the gravitation". Appl. Phys. Research, Vol. 15, No. 1: 2023

15. D. Zareski, "Confirmation of the physics unification by the ether elasticity theory", 2022 *J. Phs.: Conf. Ser*. 2197 012031

16. L. Landau and E. Lifshitz, *The Classical Theory of Fields*, (Pergamon Press, Addison-Wesley Publishing Company, Inc., Reading, Massachusetts, 1962)

17. W. F. Durand, Editor, *Aerodynamic Theory*, (New York, Dover, 1963)

18. H. A. Lorentz, *The theory of the electrons*, (Dover, New York, 1952), first published in 1909
19. A. Einstein, "The Foundation of the General Theory of Relativity" Annalen der Physik, 49, 1916, reprinted in Ref. 8, VII, pp. 109-164, Eq. (22)
20. M. Born and E. Wolf, *Principles of optics*, (Cambridge University Press), (2006), Sec. 3.1.
21. A. Salam, J. Strathdee, Phys. Rev. D 16, 2668-2672 (1977).
22. A. O. Barut AIF. Conf. Proc. 71, pp. 73-108
23. D. Zareski, "The elasto-gravitational interpretation of the 'strong interaction'". Phys. Essays. 26, 4 (2013)
24. D. Zareski, "On general quantum mechanics", Progress in Relativity, Gravitation, Cosmology, ISBN: 978-1- 61324-811-9, 2012 Nova Sciences Publishers, Inc.
25. D. Zareski, "The electron spin as resulting from the ether elasticity". Appl. Phys. Research, Vol. 6, No. 5, 2014
26. D. Zareski, "The ether theory as implying that electromagnetism is the Newtonian approximation of general relativity". Phys. Essays 27, 4 (2014)
27. D. Zareski, "The quantum mechanics as also a case of the ether elasticity theory". Appl. Phys. Research, Vol. 6, No. 4: 2014